PSYCHOSOCIAL EFFECTS OF HAZARDOUS TOXIC WASTE DISPOSAL ON COMMUNITIES

PSYCHOSOCIAL EFFECTS OF HAZARDOUS TOXIC WASTE DISPOSAL ON COMMUNITIES

Edited by

DENNIS L. PECK

Professor of Sociology
The University of Alabama
Tuscaloosa, Alabama

With Fifteen Contributors

CHARLES C THOMAS • PUBLISHER
Springfield • Illinois • U.S.A.

Published and Distributed Throughout the World by

CHARLES C THOMAS • PUBLISHER
2600 South First Street
Springfield, Illinois 62794-9265

ISBN 0-398-05618-8

Library of Congress Catalog Card Number: 89-5174

With THOMAS BOOKS *careful attention is given to all details of manufacturing and design. It is the Publisher's desire to present books that are satisfactory as to their physical qualities and artistic possibilities and appropriate for their particular use.* THOMAS BOOKS *will be true to those laws of quality that assure a good name and good will.*

Printed in the United States of America
SC-R-3

Library of Congress Cataloging-in-Publication Data

Psychosocial effects of hazardous toxic waste disposal on communities / edited by Dennis L. Peck with fifteen contributors.
p. cm.
Includes bibliographies and index.
ISBN 0-398-05618-8
1. Hazardous wastes—Social aspects. 2. Hazardous wastes—Psychological aspects. I. Peck, Dennis L.
[DNLM: 1. Hazardous Waste—adverse effects. 2. Psychology, Social. WA 671 P974]
ID1030.P78 1989
363.72′873—dc20
DNLM/DLC
for Library of Congress 89-5174
CIP

CONTRIBUTORS

Dr. William R. Catton, Jr.: A noted theorist, ecologist, and methodologist, Dr. William R. Catton, Jr. is Professor of Sociology at Washington State University. He was a creative force behind the development of the environmental division of the American Sociological Association, of which he served as chairperson. Dr. Catton is the author of four books, including the widely acclaimed *Overshoot: The Ecological Basis of Revolutionary Change* published by the University of Illinois Press.

During the past three decades, he has published more than fifty scholarly articles and numerous technical reports on a wide range of topics relating to the effects resulting from man's utilization and exploitation of the environment. Dr. Catton is Past President of the Pacific Sociological Association, is a member on the Editorial Boards of several learned journals, and serves as a consultant to state and federal government agencies.

Dr. Michael R. Edelstein: Dr. Michael R. Edelstein is Associate Professor of Psychology in the School of Social Science and Human Services at Ramapo College of New Jersey. A social psychologist, Dr. Edelstein was awarded the Ph.D. in that area by the State University of New York at Buffalo. His research of community and psychosocial effects of toxic exposure led to the recently published book entitled *Inverting the American Dream: The Social and Psychological Dynamics of Residential Toxic Exposure* (Westview Press, 1987). He is currently a member of the Center for Social Impact Assessment team charged with the responsibility of assessing high level nuclear waste siting and evaluating the Three Mile Island Emergency Response Plan.

Dr. Edelstein's interests in applied social psychology and environment have culminated in numerous professional papers published in a wide range of interdisciplinary journals, especially in the areas of social impact assessment, environmental systems, toxic exposure, and disaster. His interest in radon gas exposure recently resulted in a foundation grant for research in this area.

Dr. Charles E. Faupel: Dr. Faupel, Assistant Professor of Sociology at Auburn University, has numerous professional affiliations including membership in the Committee on Disaster Research, International Sociological Association. A graduate of the University of Delaware, Dr. Faupel's interests in law, environment, and disaster research have led to scholarly articles and publications, including *The Ecology of Disaster* (1985) and the coauthored *Disaster Beliefs and Emergency Planning* (1985). He has also authored technical reports for such prestigious groups as the National Academy of Sciences and is a contributing author to edited books.

Dr. Kurt Finsterbusch: Dr. Kurt Finsterbusch is Associate Professor of Sociology at the University of Maryland where he has been a member of the faculty since 1972. He is active in professional associations including the International Association of Impact Assessment, Society for International Development, World Futures Society, Rural Sociology Society, and the American Sociological Association.

The author/editor of numerous books and readers, book chapters, and journal articles, these publications focus on environmental and related social issues. Included are *The Methodology of Social Impact Assessment; Social Research for Policy Decisions, and Understanding Social Impacts.* Dr. Finsterbusch's research articles have appeared in a variety of sociology and interdisciplinary journals such as *The Journal of Conflict Resolution; Studies in Comparative International Development; Social Problems, Journal of the International Society for Technology Assessment; Society; Social Science Research; Sociological Methods and Research; Pacific Sociological Review; Urban Analysis; Environment and Behavior; Sociological Inquiry; and Technology Forecasting and Social Change.*

Dr. Duane A. Gill: Currently Assistant Professor of Sociology at Mississippi State University, Dr. Gill previously served as Research Associate, Marine Recreation Laboratory, Department of Recreation and Parks at Texas A & M University. Dr. Gill's research and teaching interests include social impact assessment, environmental sociology, human ecology, population analysis, and social change. A graduate of Texas A & M, he wrote a dissertation entitled *A Disaster Impact Assessment Model: An Empirical Study of a Technological Disaster* as a partial requirement for the Ph.D.

Dr. Gill has coauthored several journal articles, monographs, book chapters, and technical reports. He is active in professional programs through presenting research papers and serving as a member of various panels.

Mr. William K. Hallman: Mr. Hallman's activity and expertise in evaluation, computers, and environmental issues and policy are already established. He is currently serving as Information Resource Consultant with the State of South Carolina State Health and Human Services Finance Commission and is a candidate for the Ph.D. in social psychology at The University of South Carolina, Columbia.

In addition to these efforts, Mr. Hallman is an active consultant and researcher whose work is supported through grants from the National Institutes of Mental Health, the United States Department of Health and Human Services, and the American Cancer Society. He is the author of numerous papers and technical reports in the areas of needs assessment, computers, responses to hazardous toxic waste, and community service delivery.

Dr. Allan C. Mazur: Dr. Allen C. Mazur is Associate Professor of Social Science in the Maxwell School at Syracuse University. In addition to earning a Ph.D. in Social Relations from Johns Hopkins University, Dr. Mazur has a M.S. degree in engineering from UCLA and a B.S. degree in physics from IIT in Chicago. His major interests and areas of research include technology, biosociology, and research methods. The author of numerous research articles, Dr. Mazur has published the results of his research in wide-ranging, interdisciplinary journals on topics such as political values, environmental controversies, public opinion and public reaction, aggressive behavior, science and public policy, measurement of public opinion, ocean research, physiology, small groups research, suicide, nuclear waste, technological innovations, and conflict resolution.

Mr. Michael W. Mullen: Mr. Michael W. Mullen is Research Coordinator and Assistant to the Director for Environmental Research and Service at Troy State University in Alabama. His work at the Center has involved research on wastewater finance including state revolving loan funds and privatization, studies on improving environmental data and information utilization, and examination of successful rural solid waste collection models in Alabama.

In addition to his professional interest in environmental issues Mr. Mullen is active in the Sierra Club and currently serves as chairperson for the Alabama Chapter of the Sierra Club. As a member of the Sierra Club's national Hazardous Materials Committee, Mr. Mullen played a major role in developing club policy on hazardous waste incineration, and he is one of the Sierra Club's leaders in efforts to block shipboard incineration of hazardous waste.

Dr. John W. Murphy: Dr. John W. Murphy, Associate Professor of Sociology at The University of Miami, is a recipient of the Ph.D. from Ohio State University. His areas of expertise include evaluation research, theory, public policy, technology assessment, and the application of computers, law, and social organization.

Dr. Murphy previously served as Director and Program Evaluator for the Health Careers Opportunity Program at Ohio State University, and Director Research and Program Evaluation, Community Action Against Addiction in Cleveland, Ohio, and at Arkansas State University. He is the author/coauthor of five books: *Qualitative Methodology: Theory and Application* (1983); *The Social Philosophy of Martin Buber* (1982); *The Underside of High-Tech* (1986); *Technology and Human Productivity* (1986) *and Responsible Institutions: Organizations Without Control* (1986).

Dr. Murphy has published more than eighty articles in a variety of learned journals, served as theme editor for five special journal issues, and has published more than 66 book reviews. Active in several professional associations, he is also a member of the Council of Advisors Institute for Advanced Philosophic Research at the University of Colorado, and serves as Associate Editor for several scholarly journals.

Dr. Dennis L. Peck: Dr. Peck is Editor of *Sociological Inquiry,* the International Sociology Honor Society Journal, and Professor of Sociology at The University of Alabama. A graduate of Washington State University, he is the author of more than thirty-five scholarly articles, a book, chapters, and technical reports on topics related to public policy, suicide, crime, labor relations, safety, posttraumatic stress disorders, education, and population. Dr. Peck is a Past President of the Mid-South Sociological Association, is an Editorial Board member of several journals, and he served as Senior Analyst with the U. S. Department of Education and the U. S. Department of Housing and Urban Development in Washington, D. C.

Dr. J. Steven Picou: Formerly Professor of Sociology at Texas A & M University, and now Chair of the Sociology, Anthropology and Social Work Department at the University of South Alabama. Dr. J. Steven Picou is well known for his application of the sociological method in social impact assessment and the legal system, organization theory and policy, social change, and public policy analysis. His expertise as an evaluation researcher has led to appointments as consultant to private business corporations, state and federal government agencies, including the Central Texas Health Systems Agency, the National Science Foundation, and the National Institute of Education.

Also a professionally active scholar, Dr. Picou is the author of two books, twelve research monographs, and eight book chapters focusing on youth, career development, and the sociology of sport. In addition, he has authored or coauthored more than forty journal articles as well as numerous book reviews. His current efforts include a book in progress, *A Sociology of Technological Disaster: Dimensions of a Modern Social Problem* (with Duane Gill) and research in human ecology, environmental impact assessment, and technological disasters.

Ms. Alice A. Rahill: Ms. Alice A. Rahill is past Director of the Post Script Consulting Service in Corning, New York and is currently engaged in graduate clinical study at Wayne State University. Previously Ms. Rahill earned a Master of Arts degree from Alfred University, specializing in school psychology.

A background as school psychologist and teacher at both the secondary and postsecondary levels, as board member of community organizations, and her clinical training have led Ms. Rahill to develop research interests in the effects of hazardous toxic waste and quality of community life factors. Ms. Rahill has authored numerous conference papers which focus on learning disabilities and gifted children, and she is an active member of several professional societies.

Dr. Richard Walker: A policy analyst with the State of Illinois Department of Nuclear Safety, Dr. Richard Walker is one of an increasing number of former academics who apply rigorous technique and methodologies to environmental problems. His current interest include the synthesis of psychoanalytic concepts and community responses to social problems.

Dr. Abraham Wandersman: A Professor of Psychology at the University of South Carolina, Dr. Wandersman interests lie in community perceptions of risk, citizen participation, and social networks. He is an active scholar whose articles have been published in a number of interdisciplinary journals. Dr. Wandersman has authored numerous book chapters and was a contributor and coeditor for *Beyond the Individual: Environmental Approaches and Prevention* (1985), *Neighborhood and Community Environments* (1987), and he is coauthor of the well-received *Psychology and Community Change.*

Dr. Wandersman's research interests include the personality, community participation, community development, evaluation, public policy, and the environment; areas in which he has been a major contributor to the learned literature.

Dr. Rae Zimmerman: An Associate Professor of Planning at New York University, Dr. Zimmerman has broad research and teaching background in the areas of environmental management and planning, risk assessment, energy resources, urban economics, and organizational theory. She has published numerous articles in learned journals as well as monographs and chapters on such topics as hazardous waste management policies and implementation, environmental policy, and risk assessment, pollution control effectiveness, evaluation of public attitudes, organizational evaluation, and government regulations. A consultant to state and federal governments and an active participant in public service, Dr. Zimmerman combines a unique ability to translate academic research and writing with the practical need to move toward solutions of industrial and environmental problems.

Introduction

PSYCHOSOCIAL EFFECTS OF HAZARDOUS TOXIC WASTES ON COMMUNITIES

DENNIS L. PECK

There is much interest and concern for the consequences related to toxic chemical contamination. Some of this is expressed through learned analyses of decision-making and risk (Clarke, 1988), while others demonstrate a similar interest and concern through social activism. A culture of silence in which issues related to public safety are ignored appears to be a part of the past.

Since publication of Levine's *Love Canal: Science, Politics, and People* (1982), both the public and politicals are increasingly sensitive to the inadequacies of hazardous waste management practices and the sites selected for the disposal and storage of hazardous waste materials, of which a large portion is toxic. Some analysts argue that dangerous conditions (e.g., Bogard, 1988), such as toxic contamination episodes create the conditions leading to community action. This kind of reaction occurs, according to Edelstein and Wandersman (1987, pp. 105–106)

> . . . because environmental turbulance demands efforts to create stability and because the residents are isolated from sources of support previously assumed to be available.

In another more intriguing discussion, however, Krajick (1986) suggests that risk homeostasis may also operate where safety is concerned. Quoting one safety expert. Krajick writes that over time

> . . . people will always adjust their behavior in such a way as to offset the overall effects of any safety improvement. . . . Every society has a built-in target level of risk that constitutes the level of danger people are willing to accept in exchange for the benefits they believe they accrue. . . . This target level functions as a kind of danger thermostat (1986, p. 10).

Krajick's discussion refers to traffic safety and Gerald Wilde's theory of risk homeostasis which, according to Krajick, may explain why devices that make driving safer may also cause a collective increase in risky behavior. Risk homeostasis represents an important concept which may be useful to explain why current public policies do not seem to be responsive to concerns about waste management practices. This may occur because some "... people adjust their behavior even in response to changes in the environment that are not readily apparent ..." (Krajick, 1986, p. 11).

Risk homeostasis may be related in part to the risk-benefit perspective in which it is assumed that available options have some good features (e.g., financial profit) which contribute to utility, and some bad features (e.g., pollution and health risks), which detract from utility. A crucial assumption is that positives can and should be balanced against the negatives. In this kind of risk-benefit evaluation, however, amoral and value-free judgments are difficult at best. According to Berk and Rossi (1976), evaluation research designs, which include value judgments, are inevitable. This subjective assessment among evaluation researchers is no less true for the contributors to this volume.

One purpose of this volume is to demonstrate that the problems related to hazardous toxic waste disposal affect real people and that the effects of improper disposal practices have long-term consequences. As Restivo (1988, p. 206) points out, modern science is widely assumed to be a successful and progressive enterprise. But Restivo also argues that science may itself be considered a social problem. Our nation faces the consequences of scientific and technological progress, both positive and negative. The communities where we live and work experience the direct effect of hazardous waste materials, including toxic chemicals, in a multitude of forms. In addition to managing the transport and disposal of these wastes, professionals must deal with the psychosocial effects that the mere presence of toxic substances may have on communities.

This volume brings together experts who address contemporary issues and offer a sense of the direction of future public policies. These issues are vital to the environment and all living things, and include

- the psychosocial effects of scientific and technological progress on our communities;
- the responsibility for public health in environmental issues such as facility siting;
- legislative guidelines for processing hazardous toxic waste, and the problems of human exposure or victimization;

- community preparedness for potential emergence, and the psychosocial effects posed by hazardous toxic waste management, transport, and disposal; and
- future directions for hazardous waste management and its psychosocial effect on communities.

This is not a book about theory, nor is it intended to test theory. It is not intended as a statement of evaluation research or impact assessment exercises. The book is an issues-oriented compendium; its chapters refer to the social implications of prolifering technology, technological disasters, tragedies related to hazardous toxic waste episodes, community reactions, liability, risk assessment, safeguards for the disposal of hazardous toxic waste, monitoring of waste sites, community preparedness, public policy, and environmental planning. What the reader will find are chapters, written by acknowledged experts, in areas such as environmental sociology, impact assessment, toxic waste management, and disaster. The authors of these chapters have written insightful statements. We hope the readers will agree with this assessment.

Progress is inevitable, but history demonstrates also that this inevitability is responsible for both positive and negative consequences (Catton, 1982). Most of the chapters in this publication address the deleterious effects of indiscriminate or unwise disposal of hazardous toxic waste materials. These effects range from physiological damage suffered by individuals to the contamination of entire communities. Other chapters address legal and policy issues. You will be interested in reading this collection of papers if you represent the hazardous waste industry, a municipal, state, or federal government agency, engage in public policy and legislative activities relating to the hazardous waste industry, or participate in human service activities. The information will be useful to industrialists, educators, the legal profession, and public health practitioners interested in the effects of scientific and technological progress, toxic tort law, current guidelines for processing hazardous toxic waste, human exposure assessment, victimization, preparedness for emergencies, the results of disaster research, and suggested guidelines for the future. A brief summary of each chapter follows.

With a focus on the conflicting definitions of risk assessment, Michael W. Mullen addresses issues relating to this concept within the public and private sectors. The discussion highlights efforts to repress or diminish public concern about hazardous waste management procedures. A series of questions about risk assessment are presented, and these questions then are used to establish the important link between a theoretical concept and public policy decisions.

The chapter entitled "The Effects of Prenatal Environmental Toxin Exposure: A Case Study" is developed using a case study of a female child whose mother had been exposed to environmental toxins, including dioxin, during the first three months of pregnancy. Using medical records and the results of achievement scores Ms. Alice Rahill first describes and then evaluates the physiological, learning, and social disabilities the child experienced during childhood and early adolescence. Complete with an extensive bibliography, figures, and tables, this chapter moves beyond the specifics of a case study to illustrate the need to evaluate the effects of toxins on learning skills and behavior of children.

In a chapter entitled "Perception of Risk and Toxic Hazards" William Hallman and Abraham Wandersman have created a statement to explain the diversity in public perceptions of and opinions toward hazardous toxic waste and current efforts to deal with this material. This discussion covers a number of issues of importance for understanding the scope of the problems associated with real and perceived psychosocial effects on communities. Included in this discussion are analyses of stress, coping, community resources, individual and family outcomes, group and community organization, and future implications for effective risk communication and research.

"Community Responses to Exposure to Hazardous Wastes" is a well-documented, thorough chapter representing original research. Dr. Finsterbusch's chapter first summarizes the extensive knowledge of case studies of community responses to exposure to hazardous waste and then relates these findings to three areas; namely social movements, community conflicts, and the political economy of environmental problems.

In the second part of this chapter the author discusses the general pattern of toxic waste contamination episodes using these patterns to illustrate the case of a large hazardous waste site. In the final section the author provides an indepth description of the behavior of the victims of 25 toxic waste contamination episodes. Specific cases are then selected for more detailed analysis.

In "Toxic Waste Disposal Sites as Technological Disasters" the authors highlight the magnitude of toxic waste production in American society as well as the distribution of this product. Framed in a sociopolitical discussion, Drs. Gill and Picou provide a significant review of the literature pertinent to technological disaster and its consequences, cite specific examples, and then assess the importance of Superfund legislation for controlling future disasters. The final section of this chapter includes issues relating to current and future social policy, research needs, and relevant sociological issues.

William R. Catton, Jr.'s vivid description of Times Beach, a small town no longer listed on road maps, dramatizes the disasterous effect of a hazardous waste site contaminated with dioxin. The reader is informed of the woes that befell Times Beach, Missouri, including the physiological disorders that afflicted residents leading eventually to the closing of the entire community. This intriguing discussion also highlights several examples of community reactions which demonstrate a continuing ignorance of the devastating effects of hazardous toxic waste disposal on communities. The remainder of this chapter evaluates the historical and sociological relationships between Homo sapiens and the physical environment.

The focus of the next chapter, written by Dr. Allan Mazur, is the quality of news media coverage of hazardous waste sites. Based on the authors' experience and expertise in assisting agencies to shape public opinion, Mazur's effort is focused on the techniques employed by the news media to disseminate information about hazardous toxic waste issues to the public. This issue is particularly germane to the psychosocial effects of hazardous toxic waste problems because of the publics' need to be informed of current efforts within both the government and the private sectors to deal with the risks involved.

Drs. Charles E. Faupel and Conner Bailey begin their discussion with a brief review of the events leading to the passage of federal legislation such as the Solid Waste Disposal Act and the Resource Conservation and Recovery Act. They then discuss the potential hazards to human health and the environment and the potentially catastrophic risk involving the concentration of large amounts of hazardous materials in central locations.

Assuming that hazardous waste emergencies will occur, the authors assess the importance of careful contingency planning. Basing their argument on a number of principles, which are derived from several decades of natural disaster research, Faupel and Conner address these problem areas by drawing upon observations of the nation's largest hazardous waste landfill. Using information gleaned from interviews conducted with representatives of emergency organizations and other knowledgeable persons, this chapter is directed toward emergency planning and preparedness. This is a process-oriented description of local politics, economic and community life in a county which is financially dependent on the by-product of advanced technology; namely, the hazardous waste industry.

Michael R. Edelstein's research and experience as an expert witness hired to provide court testimony serve as the basis for his chapter entitled "Psychosocial Impacts on Trial: The Case of Hazardous Waste Disposal." In this well-written chapter, Dr. Edelstein provides important insights regarding the litigation over the psychosocial effects of toxic hazardous waste. Edelstein's case study analysis is based on the participant observation approach and his analysis of litigious actions and toxic torts are both insightful and useful for understanding the legal implications related to psychosocial impact assessment.

"Characteristics of Citizen Groups Which Emerge with Respect to Hazardous Waste Sites," written by Dr. E. L. Quarantelli, is a sociological discussion of citizen groups which are formed to prevent or prepare for disasters or to cope with the effects of community disasters. The author draws upon a great experience in the area of disaster research and his expert analysis of the emergence, organization, and strategies of private citizen groups is a noteworthy contribution to any discussion of the effects of hazardous waste on communities.

In "Public Acceptability of Alternative Hazardous Waste Management Services" Dr. Rae Zimmerman discusses another in a series of important issues and problems which this volume is intended to address. The author's discussion of the scale and distribution of hazardous waste sites represents a learned and practical assessment of sensitive issues related to hazardous waste management, protection of the public and the environment, and the change which has recently occurred in the handling of hazardous waste.

Dr. Richard Walker's application of psychoanalytic theory to the problem of siting hazardous toxic waste facilities is intended to stimulate thought and to upset prejudices. The author's evaluation of community reaction to hazardous toxic waste and the siting of this material is an intriguing explanation based on Freudian concepts. This chapter is a well-documented exposition of society's aversion to the waste and siting issues. It is also complete with current data and other specifics pertaining to sites, the economic costs of managing hazardous sites, and current policies.

Dr. John W. Murphy assesses the social implications of a proliferating technology as this phenomenon pertains to the disposal of hazardous toxic waste. Utilizing a systems approach to evaluate a cycle of self-perpetuating technology, Murphy's argument is contrary to the popularly held view that technology serves as a tool to solve social problems and to stimulate new environmental policies. Combining a postmodernism discussion with the hazardous toxic waste problem, which is central to this volume, Murphy's assessment of the technological worldview is as stimulating as it is challenging of current knowledge of and orientations toward the use of new technologies to solve problems.

REFERENCES

1. Berk, Richard A., and Rossi, Peter H.: Doing good or worse: evaluation research politically re-examined. *Social Problems, 23:* 337–349, 1976.
2. Bogard, William C.: Bringing social theory to hazardas research: conditions and consequences of the mitigation of environmental hazards. *Sociological Perspectives, 31:* 147–168, 1988.
3. Catton, William R. Jr.: *Overshoot: The Ecological Basis of Revolutionary Change.* Urbana, University of Illinois Press, 1982.
4. Clarke, Lee.: Explaining choices among technological risks. *Social Problems, 35:* 22–35, 1988.
5. Edelstein, Michael R. and Wandersman, Abraham.: Community dynamics in coping with toxic contaminants. In Altman, I. and Wandersman, Abraham.: *Neighborhood and Community Environments.* New York, Plenum, 1987, pp. 69–112.
6. Krajick, Kevin.: Do seat belts kill? *Psychology Today* May: 10–11, 1986.
7. Levine, Adeline G.: *Love Canal: Science, Politics, and People.* Lexington, D. C. Heath, 1982.
8. Restivo, Sal.: Modern science as a social problem. *Social Problems, 35:* 206–225.

ACKNOWLEDGMENTS

This book is dedicated to Lakshmi K. Bharadwaj, teacher and friend, whose support and encouragement provided a young graduate student direction and nurtured the desire to successfully complete a masters level program, and later to pursue a course of study at the doctoral level. His compassion, strong resolve, and character continue to serve as a personal source of inspiration.

The idea for this volume emerged out of discussions held with Raymond M. Hollub, Charles E. Forman, Bryan C. Graham, Dennis P. Prisk, and others during a year and a half period. These discussions became a part of a planned project of which several of the contributors to this volume agreed to participate. I appreciate the gracious responses and generous efforts given so freely by the authors who agreed to contribute to this effort.

My wife, Sharon, and daughters, Heather Kathleen and Shannon Marie, provided their usual tolerance for my work habits which keep me in the office late into the evening and often during the weekend. A special appreciation is acknowledged to Kimberly A. Folse for her support on this project and others too numerous to count. And a special thank you is given to Sandra Arnold for her patient assistance in typing and otherwise handling communique related to this project.

CONTENTS

PSYCHOSOCIAL EFFECTS OF HAZARDOUS TOXIC WASTE DISPOSAL ON COMMUNITIES

Chapter 1

THE ROLE OF RISK ASSESSMENT AND COMMUNICATION IN COMMUNITY RESPONSES TO HAZARDOUS WASTE MANAGEMENT PROJECTS: POTENTIAL ABUSES OF RISK ASSESSMENT

MICHAEL WILLIAM MULLEN

Risk is a term that apparently means entirely different things to different people. Even those involved in professional risk assessment work appear confused about the meaning of risk. According to Shistar (1988), the Definitions Committee of the Society of Risk Analysis after two years of study has been unable to select a definition of risk from thirteen under consideration. As it is used by risk assessors and bureaucrats, risk is a statistical measure of the probability (and perhaps the severity) of adverse effects. Thus, in the context of risk assessment, risk is generally considered much more narrowly than the public generally thinks of it.

Risk can apply to a group such as consumers of a contaminated water source, workers in an industry or it can be applied to individuals. Also, risks can be grouped into voluntary and imposed risks. Voluntary risks are generally not regulated. For example, excessive intake of salt increases one's risk for coronary disease and strokes but is not regulated. It is generally imposed risks which are regulated as a means of exercising collective choice (as opposed to individual choice) in minimizing risk.

Risk assessment generally attempts to estimate the chance that an average person in the affected population will be injured in a particular manner—such as developing cancer. If risk assessment can accurately estimate health risks to populations from various activities or chemicals, the technique can be extremely valuable to society. Since many vital activities of modern society involve risk, and since eliminating risk more frequently than not involves a significant commitment of resources, accurate estimation of various risks faced by society can allow allocation of resources for efficient reduction of total risk to populations. Therefore, risk assessment can be (at least theoretically) a very beneficial tool for society.

Unfortunately, risk assessment can and is being utilized by bureaucrats and by industry to justify "What, me worry?" attitudes about risks rather than to examine how imposed risks can be reduced and which risks society should work to reduce first. Well known and respected scientists, including Ames (1987), Ames, Magaw, and Gold (1987), Tierney (1988), and Bennetta and Winton (1987), have published materials which seek to minimize the significance of imposed and often avoidable risks by pointing out that natural toxins in foods or other voluntarily assumed risks are larger than various imposed risks. The confusion created by such comparisons, which are actually comparisons of apples and oranges, is detrimental to society since it delays or prevents formation of public consensus on the real issue—*reduction of imposed risks.* Risk assessment in such instances appears very similar to past attempts by bureaucrats and industry to parade scientists before the public telling them "everything is safe," "not to worry," "trust us." Fortunately, from the perspective of reducing risk, the public is not easily deceived by these attempts to use risk assessment as a substitute for public policy formulation.

Risk assessment and communication can clearly be either a tool for improving public understanding and involvement in complex societal decisions or an impediment which can further erode public confidence in regulatory agencies. If agencies which do risk assessments try to use them without a public policy foundation to justify their actions, risk assessment and communication will do little to solve society's waste management or other problems to which risk is inherent. Instead, if these agencies recognize that the public must make the choices as William Ruckelshaus(1986) did in the following comments at a national conference on risk assessment, development of public policy can be facilitated by risk assessment and communication techniques.

> ... the question before us is not whether there is going to be a sharing, whether we will have participatory democracy with regard to the management of risk, but how. How do we make decisions effectively, efficiently, and in a timely way, keeping the quality of the decisions high and at the same time involving the public?
> Now, none of this will work unless we figure out how we are going to effectively involve the public.... Certain elements need to be present

> for participatory democracy or public involvement to work. The first is access to the decision maker. . . .
> The second principle,. . . . , is that before access can become meaningful, there must be sufficient information in the hands of the public for citizens to be effective participants. . . .
> The information that is provided by the decision maker has to be honest, clear, and as complete as we know how to make it. . . . If we don't know something, we must say we don't know, and if we are uncertain, we must say we are uncertain. . . . Where risk communication usually starts to break down is after the decision is made and there is a perceived necessity to sell a very uncertain, shaky judgement to the American people. . . . Unless we inculcate very deeply in our risk management agencies a real and perceived honesty in communication, we are going to lose the opportunity to bring the public along in understanding and in responding sensibly to the mix of issues that need to be weighed."

Unfortunately, not all who would apply risk assessment seem to share the views expressed by William Ruckelshaus. There are some who do not seem to share the view of decisionmaking on societal issues involving risk. Far too many seem to want to revert to a by-gone era when agencies put a "scientist in a labcoat out on the stage" to tell a trusting, gullible public not to worry. The only difference seems to be that now the scientist will say don't worry, the imposed risk that you will be asked to accept so that the company down the block can profit is smaller than risks you already accept voluntarily. The difference is that individuals in the public receive something from that voluntary risk (real or perceived pleasure, and happiness) and they individually may not realize or at least fail to recognize any benefit from that proposed waste disposal facility down the road.

Too many people in industry and research have used and abused this comparison between imposed and voluntary risks. An example is a table which Ames, Magaw, and Gold(1987) used showing the relative risks of various chemicals, foods and activities to the risk from a glass of beer or wine. It is not valid to compare the risk from a glass of beer or a peanut butter sandwich from which the consumer draws some immediate personal satisfaction and benefit to the risk from an emission from a proposed waste treatment facility which is to be located down the road. This is an abuse of risk assessment and communication. It ignores the fact that imposed and voluntary risks are in their very nature different. Even if it were not somewhat intellectually dishonest, it would still be counterproductive to use such comparisons. Common sense on the part of the public rejects this type of risk comparison.

Since risk assessment can be readily abused and in some cases is already being abused by bureaucrats and industry representatives, the public is understandably distrustful of risk communication efforts. Risk assessment and communication cannot be successfully used to hide or camouflage public policy decisions made in a secretive manner by bureaucrats. The risk assessment and communication process can not and should not be used to try to minimize the public's concerns about imposed risks by comparing smaller imposed risks with larger voluntary risks. Instead, both classes of risks should be distinguished so that individuals can make personal choices to reduce voluntary risks and decision-making units in society can make collective decisions leading to appropriate reductions in imposed risks. Risk assessment will not solve the waste facility siting crisis by simply telling the public that imposed risks from toxic chemical exposure are smaller than voluntarily assumed risks like smoking, drinking, or dietary choices.

Certainly the public needs to be informed about risks associated with hazardous waste management. However, the risks which need to be compared are those for different waste management or site cleanup strategies so that a meaningful comparison of the costs and risks for various options can be made. Going beyond end-of-pipe waste management, risk assessment should be utilized in cost benefit analyses of new processes and/or products which result in the generation of hazardous wastes. Thereby, society can collectively choose products which minimize waste production and waste management policies which reduce risk from necessary waste management activities while conserving scarce technical and financial resources. Only in this manner can local communities seek waste site operations which maximize risk reduction within regulatory or economic constraints.

Since risk assessment is not a purely scientific exercise and is fraught with assumptions which should only come from public policy decisions, communities and individuals should carefully scrutinize all risk assessments. Several important questions have been developed by the Sierra Club's Risk Task Force (Shistar, 1988) as follows:

I. THE CONTEXT OF THE RISK ASSESSMENT

- Why is the risk assessment being performed? To set standards? If so, are they standards for maximum contamination or minimum cleanup? To set an "acceptable risk" level? To set priorities? For a risk-benefit analysis? Does the risk assessment take the place of more protective measures? If so, are the more protective measures (e.g., zero risk) feasible?
- Who pays the cost of reducing risk?
- Who benefits from reducing the risk? Is one segment of the population affected more than others?
- Who benefits from the process that imposes the risks?
- Is the risk imposed voluntarily?
- To what extent is the process that produces the risk an essential activity? Are there alternatives? Are the alternatives assessed?

II. HAZARD IDENTIFICATION

- Which risks are assessed? To whom or what?
- How does the risk assessment deal with uncertainty?
- Are all sources of risk from the process or material assessed (manufacture, use, transportation, storage, disposal)?

III. DOSE-RESPONSE

- How does the risk assessment deal with uncertainty?
- How does the risk assessment deal with variation in response?
- Does the assessment take into account the latency period following exposure?
- Are statistical inferences valid? Are questions phrased properly? Are null hypotheses protective?
- What models are used for extrapolation? Are they protective? (That is, do they err on the side of false positives rather than false negatives?) Is a threshold assumed?

IV. EXPOSURE ASSESSMENT

- How is uncertainty dealt with?
- How is variation to exposure handled? Has proper account been taken of the age distribution and location of segments of the target population in relation to the source?
- Is an environmental fate model used to identify all routes of exposure to materials, degradation products, and all steps of the process?

- Has proper account been taken of the variation in the amount of pollutants emitted over the period during which observations have been made?

V. RISK CHARACTERIZATION

- What units are used to express the risks? If risks are compared, is there a common measure of risk?
- Is a measure of the uncertainty of the risk presented?
- How is the variation in risk expressed?
- Has proper account been taken of the latency period for developing certain adverse effects?
- Is a finding of "no adverse impacts" based on insufficient data?
- Does the risk assessment have a tendency towards false positives or false negatives?
- Do different investigators disagree on the risk assessment either because they employ different data sets or different hypotheses or methodologies?
- Is the risk characterized as a population risk or individual risk? Is the characterization appropriate?
- Does the risk assessment take into account external sources of risk that may act as synergists?
- Are assumptions made explicit?
- How are different risks combined?

The questions presented concerning the context of the risk assessment are perhaps the most important to the public. If the context is inappropriate the risk assessment will not only fail to aid in consensus building, it will detract from any consensus building effort. Too many times corporate public relations personnel and bureaucrats have tried to illustrate the risk of an activity by comparing it with the risk of a dietary habit such as salt intake, driving, or smoking. The public (at least the vocal activist segment) will not accept such comparisons. The public wants and deserves meaningful comparisons of risk. For instance, the public wants risk comparisons for different technologies or policies for waste management and not comparisons of risks imposed by facilities and voluntary risks.

The public also wants risk assessment to be used for risk reduction and not for defense of risks imposed by facilities which may not directly benefit them. The issue of fairness arises when facilities are sited predominately in poor rural communities in which people benefit little from the materials whose production results in waste production and risks which they bear disproportionately.

The second category of questions concerns whether the risk assessment addresses the right target or the targets really at risk. The problem can be illustrated by considering shipboard incineration of hazardous waste. The U.S. Environmental Protection Agency at one point tried to sell shipboard incineration to the public by comparing human health risks from stationary land-based incinerators with human health risks from shipboard incineration. One error in that approach was that the ecological risks were not compared. Another error was that risks from related activities (port area waste storage, transfer operations and transportation) did not appear to be adequately addressed by EPA. Finally, low probability events such as major hazardous waste spills were essentially dismissed even though a single major spill of the wastes (polychlorinated biphenyls) could destroy or render useless the fisheries of a major estuary.

The third set of questions deals with scientific inadequacies or uncertainties. Fundamental questions remain about whether there is a threshold concentration of the toxic substance producing the risk below which no harmful effects occur. Risk assessments must not attempt to deny such uncertainties. Instead they should seek to explain the uncertainties by generating risk estimates based on both a zero-concentration threshold and on some threshold below which effects are not expected (if data to suggest a nonzero threshold exists). Since assessments are generally going to be made based on what most people would call incomplete data, uncertainties will always exist. Assumptions, statistical analyses, and hypotheses should always be protective. That is, they should be biased in favor of greater protection. The more uncertain the data, the more assumptions should be biases in favor of protection.

The public does not want to be told that a risk assessment indicates that a project or product is worth the risk associated with it and will only cause a few excess cancers. The public wants to know how the risk of the proposed project or product compares with project alternatives or alternatives. Also, activists want to know if the benefits of a product which results in imposed risks exceed the costs. A current example is the continued use of the pesticide Alar (tradename) on apples. An imposed risk is placed on consumers of apples since they do not know which apples have been treated. Yet, consumers have not been given a choice of whether they are willing to accept apples which do not have a "perfect" appearance or to pay slightly more for apples in order to reduce the risk from this potent carcinogenic substance. Risk assessments cannot substitute for the public debate over such choices. Simply showing that a risk is small in no way substitutes for the public policy debate needed to establish public policy. It can't and it won't!

The last two sets of questions on exposure assessment and risk characterization deal with technical limitations of risk assessment. They basically deal with the question of whether the risk assessment is comprehensive enough. Have any sources of risk been overlooked or underestimated? For instance are degradation products of a waste or pesticide more toxic than the parent compound? People are also concerned about professional differences in opinion which result in different risk assessment results. Finally, individuals are less interested in risks stated in terms of excess cancers or deaths per 100,000 population. They want to know what their risk or their child's risk is from the activity. Citizens also want to know how imposed risks combine with other risks, both self-imposed and those imposed by society. For instance, it is inconsistent for companies to inform workers that smoking increases the health risks due to job-related chemical exposure and not inform the surrounding community of synergistic actions between pollutants.

In summary, risk assessment is not a purely scientific endeavor. Risk assessment relies upon assumptions which often are not even based on reasonably conclusive agreement among scientists much less upon policy decisions resulting from public debate, discussion and consensus building. Risk management can be a valuable tool if applied to public policy development on hazardous waste management and other issues involving involuntarily assumed (imposed) risks. However, risk assessment must remain a tool; it cannot substitute for public policy development and implementation. Risk assessment cannot replace the dynamic processes of individual, community and national consensus building and policy development on waste management and other issues. If bureaucrats and industry representatives continue to abuse risk assessment (either purposefully or unknowingly), mistrust of industry and the regulatory community by the public will continue to grow. The public must receive all available information and play a continuous role in regulatory decisions involving facility siting. If public involvement is not facilitated by regulatory agencies and if concerned citizens do not feel that they are part of the decision-making process needed waste management facilities, cleanup activities and other actions will continue to face delays and avoidable cost escalation. Risk assessment and risk communication cannot replace the need for public consensus building on waste management and other issues. These technologies can be a valuable tool so long as they are used honestly and are not used as substitutes for truly open public policy development. Risk assessment and communication can either be a valuable tool for society or another part of our toxic dilemma.

REFERENCES

Ames, Bruce: Six Common Errors Relating to Environmental Pollution. *Regulatory Toxicology and Pharmacology,* 7:379–393, 1987.

Ames, Bruce, Renae Magaw, and Gold, Lois: Ranking Possible Carcinogenic Hazards. *Science,* 236:271–280, 17 April, 1987.

Bennetta, William J. and Winton, John W.: Natural Toxins May Be the Greatest Danger. *Chemical Week,* 29 April, pp. 10–11, 1987.

Ruckleshaus, William D.: Communicating about Risk. In Davies, J. Clarence, Covello, Vincent T. and Allen Frederick W. (Eds.): *Risk Communication,* Proceedings of the National Conference on Risk Communication, Washington, D.C., January 29–31, pp. 3–9, 1986.

Shistar, Terry. Sierra Club Risk Task Force (Xeroxed materials), May, 1988.

Tierney, John. Not to Worry., *Hippocrates,* January/February, pp. 29–38, 1988.

Chapter 2

THE EFFECTS OF PRENATAL ENVIRONMENTAL TOXIN EXPOSURE (DIOXIN): A CASE STUDY

ALICE A. RAHILL

INTRODUCTION

This is the study of Lynne, a female child who was prenatally exposed to environmental toxins, including dioxin. Maternal prenatal environment has been documented to dramatically cause gross morphological birth defects after exposure to certain drugs or radiation during critical periods of development (Hicks, D'Amato and Lowe, 1959; Haymanker, 1962; Lenz, 1962; Rugh and Wanfromm, 1966). McConnell (1984) provides a clinopathic syndrome of dioxin exposure, including: acne, edema, thymus atrophy, heptomegaly, liver necrosis, and prophyria, while Lilis (1987) discusses a psycho-organic syndrome with more subtle but equally significant effects. Research with toxic solvents includes behavioral symptoms such as disturbances of intellectual skills, decreased attention and concentration, and memory problems. The idea that learning and behavior may also be altered by toxin exposure has been advanced by Spyker (1975), whose findings were supported by the work of Needleman, Gunnoe, Leviton, Reed, Peresie, Maher, and Barrett (1979), who studied the effects of children exposed to lead.

Somatic abnormalities represent the standard material for study in conventional teratology. However, teratogens may also have affinities for particular brain centers which do not reveal developmental abnormalities, such as abnormalities in behavior, until the individual matured (Nair and Dobois, 1968). Longitudinal research can be useful in determining the delayed effects of a particular prenatal influence on behavioral functions. Lynne represents one such case useful for evaluating, in an epidemiological sense, the effects found in a more general population.

BACKGROUND

Lynne was born in New York state of Caucasian parents during late 1976. The sequence of critical events began during the prenatal period. Her mother reports in sworn testimony that she stayed in the home of her parents located in Niagara Falls, New York for a week at a time at various intervals during the first trimester, both prior to and after pregnancy was confirmed. This residence was located on a swale and near a corner sewer line junction selected for Environmental Protection Agency (EPA) monitoring. The swale originated in the actual Love Canal dump site, ran under the home driveway, and had been leaching into the basement sump pump for several years.

A schematic summation of the history of the Love Canal area is provided in Figure 2-1. From the 1890s to 1940, Love Canal was a

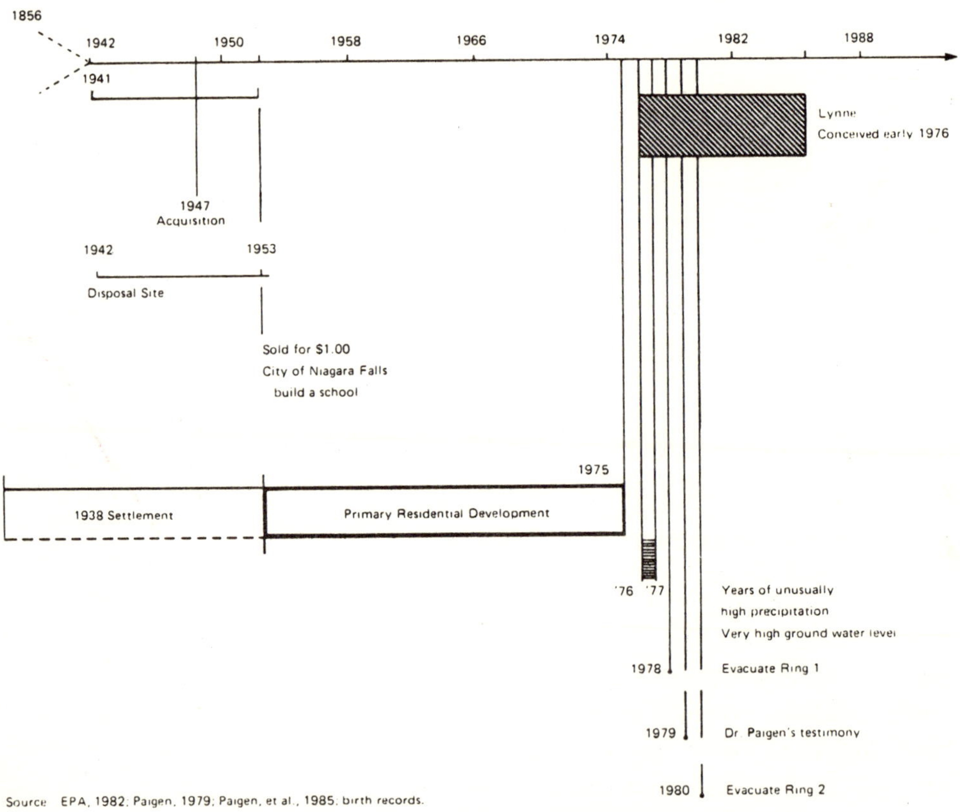

Figure 2-1. Longitudinal history of Love Canal and years of Lynne's conception and growth.

3000-meter unfinished canal. For an eleven year period from 1942 to 1953, the area was used by a local chemical company as a burial site for 19,000 metric tons of organic solvents, chlorinated hydrocarbons, acids, and other chemical wastes including dioxin (tetrachlorodibenzo-p-dioxin, TCDD). At least 248 waste chemicals were identified at the waste area (DHEW, 1980).

In the Appendix some of the chemicals found at Love Canal are listed. Theses are principally organic solvents, chlorinated hydrocarbons, and acids. Few toxicological studies have been conducted on the long-term effects of chemicals that are manufacturing by-products or intermediates, rather than commercial products. Toxicological information for 100 of the 248 chemicals was not found for the Love Canal area (Paigen, Goldman, Highland, Magnant, and Steelman, 1985). However, the Environmental Protection Agency (EPA) (1982) does confirm that varying amounts of the highly toxic compound 2,3,7,8 tetrachlorodibenzo-p-dioxin were present at some locations within the Canal area, particularly those favorable to the transport of sediment by water flow.

As indicated in Figure 2-1, Lynne's conception occured during a period of unusually high precipitation and very high ground water, 1976–1977 (EPA, 1982). In 1977, twenty-five years after the last dumping occured, chemicals were detected by public health officials in the neighborhood creeks, sanitary sewers, air, soil, and sump pumps (Leonard, Werhmary, and Zieglon, 1977). In the Appendix, a tabulation is provided of the known toxicities for the chemicals found at Love Canal.

CASE HISTORY

At age 8 years, 11 months, Lynne's mother requested an up-dated evaluation by the author, at the time a school psychologist, who had previously examined her daughter. The presenting concerns were child's behavioral pattern, which included nightly enuresis and occasional daytime lapses, poor eating habits, low weight gain, low attention span, and high activity level. The unusual pattern of Lynne's development inspired a search of the literature for a similar occurrence. Finding none, this case study was initiated. Archival data of medical and educational records were made available by Lynne's mother, and consent was given for verification from professional sources.

Birth records indicate the parents as healthy adults; Lynne's mother was 30 years of age and a nurse, her father was a 29-year-old educator.

The only previous pregnancy in the family was unremarkable. Lynne was a full-term baby; only the small size of the fetus was noted during the period of gestation.

At birth, Lynne presented a variety of defects as follows: single umbilical artery (often considered associated with serious congenital malformation); solitary left kidney (confirmed, at age 4); bilateral clinodactyly; high arched palate; hypoplastic facial skeleton; syndactyly of second and third digits of both feet; low birth weight (2310 grams); microcephalia (head circumference, 30.4 cm.); partial malformation of cartilage of outer ears; metatarsus varus.

Lynne's Agpar Scale was 8–9 (normal is equal to 7–10), therefore birth records suggest a functioning infant with multiple congenital defects. Chromosomal studies were done at birth and repeated later at different institutions. Both results were negative, as were the family medical histories. In 1981, an excerpt from Lynne's pediatric neurologist report stated, "I would suspect the influence of a teratogenic agent." (Dr. E., 2–81, medical records)

In 1979, the area of the maternal grandparent's home was determined to be contaminated with toxic wastes, including dioxin. Located in the geographic region which was to become known as Love Canal, their home was condemned and eventually purchased by the state of New York. In May, 1979, legal counsel was retained to pursue a claim on Lynne's behalf against Occidental Corporation, and others, for injuries incurred while Lynne was in utero, particularly during the first trimester of development. This claim was individually settled as part of a class action in December, 1984, and Lynne received a monetary award for damages incurred (Supreme Court, State of New York, County of Niagara, Index 36641).

Medical Development After Birth

At the age of eleven months, Lynne started experiencing absence seizures. Although developmental delays were noted in several areas, no specific remediation was suggested. Medical treatment included leg casts for metatarus varus and ventilating tubes during at age 1, 2, and 6 years due to multiple ear infections. Urinary tract infections also occured at intervals.

In 1983, photoconvulsive seizure activity prompted Lynne's neurologist to prescribe Depakene, 100mg., four times a day. Also in 1983, Lynne's high, constant activity level led to a diagnosis of hyperkinesis, overactive reflexes, and poor coordination in both gross and fine motor skills.

Ritalin was prescribed at 5mg, three times daily; medication for both conditions was monitored by a pediatric neurologist. Extensive urological studies were conducted in 1983, conclusively demonstrating the presence of a single kidney. Daytime toilet training was considered accomplished at approximately age seven, with occasional lapses.

Physical examination of Lynne at age 8 years, 2 months revealed a child who was within the lowest 3 percent of normal growth. At this time her weight was 39 pounds; height 46 inches, and head circumference 46 cm.

Academic Development

To assist her educational development, Lynne was placed in the Early Intervention Program, the Resource Room, and a Learning Disabilities Tutorial Program. Early in 1984, she was placed in an Option Four (Physically Handicapped) classroom. Considered unable to attend in a conventional class atmosphere, Lynne's classification rendered by her district's Committee on the Handicapped indicates she was "Physically handicapped—neurologically impaired."

From age 4 years 2 months to 5 years 5 months, Lynne was administered three Stanford-Binet Form L–M Tests (Terman and Merrill, 1960), by two examiners, for which she measured an intelligence quotient of 84, 89, and 89, respectively. The Wechsler Intelligence Scale for Children-Revised [WISC–R] (1974), administered by yet a third examiner at age 6 years 10 months, resulted in a Full Scale 92, Verbal 90 and Performance 92.

The evaluation requested at age 8 years 11 months was done by an examiner who had previously dealt with the child; this test yielded WISC–R scores of Full Scale 91, Verbal 90 and Performance 95. These scores indicate a general profile of ability within the normal range ("normal" defined as a score of 90 to 110), with a balance between language and visual-spatial skills. However, subtest analysis indicate Lynne's distractibility factors sum to 78, lower than normal by nearly one standard deviation from the mean, and indicative of a deficiency in this ability.

The Bender (Koppitz, 1964) profile was well organized, with two columns of designs in sequential order. Visual motor production was delayed, scoring at a Mental Age of 6 years 3 months, but designs were conceptually correct. Reading achievement was below the age-ability expectancy, (deficiency level for Lynne was grade 1.7), as measured on the individual Stanford Diagnostic Reading Test (Karlson, Madden and Gardner, 1976). Mathematical reasoning was stronger than application, which scored in

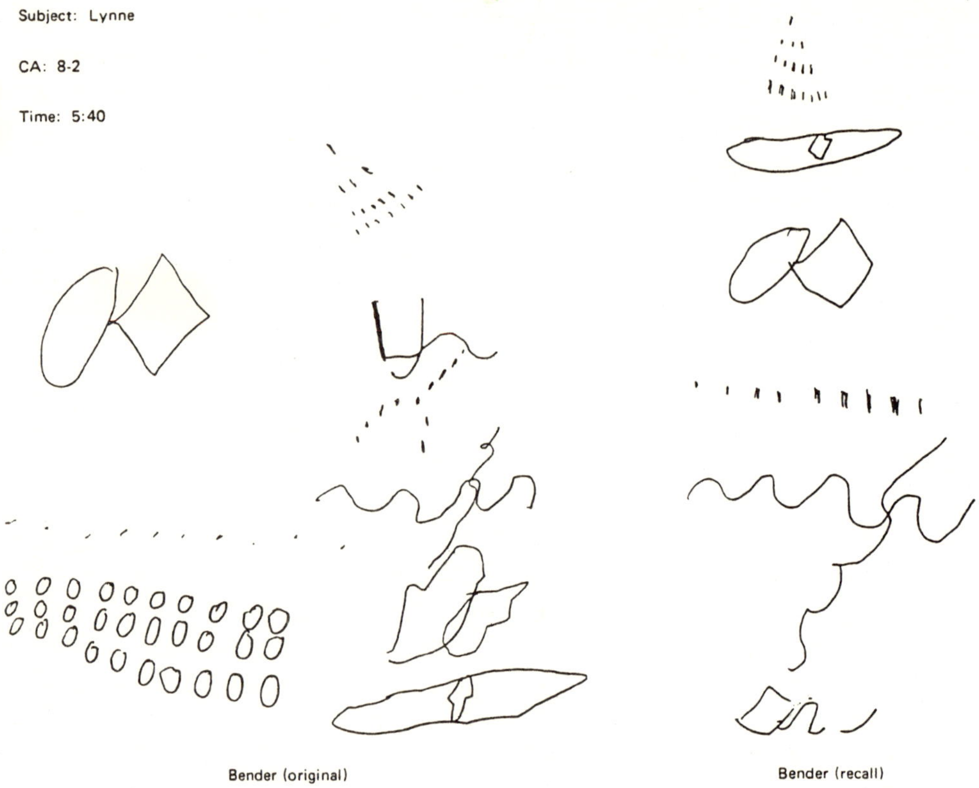

Figure 2-2. Bender Gestalt Drawing Test.

the deficiency level according to the Stanford Diagnostic Mathematics Test (Beatty, Madden, and Gardner, 1976). Daily performance and classroom achievement were limited by almost illegible handwriting and spelling. Written production is characterized by severe delays, extensive subjective effort, and minimal results. When asked to write the alphabet in order, a task which usually requires 40 seconds for an average child of this age, Lynne required 2 minutes and 20 seconds to complete this task. The result was a mixture of upper and lower case forms.

When asked to write a paragraph of her choice, the result was:

"the bers
p. m. b.
the bers head
the sorb
hthe woke
woka
bed."

The story of the three bears was verbally related in much greater detail. However, verbal expression was characterized by difficulties of word retrieval, syntax, and fluency. The Peabody Picture Vocabulary Test (Dunn and Dunn, 1981) is intended to measure verbal receptive language. On form A of this instrument, Lynne measured an IQ of 115, with a mental age projection of 12 years, 4 months. The Expressive One Word Picture Vocabulary Test (Gardner, 1982) presents a picture for the subject to name. This measure yielded a mental age score of 9 years 7 months. The different scores between these tests indicated that Lynne had a higher receptive language knowledge than was suggested by her expressive language. Subtests of the Illinois Test of Psycholinguistic Abilities (Kirk, McCarthy, and Kirk, 1968) measured a mental age of 8 years 2 months for both Auditory Closure and Grammatic Closure. Lynne scored in the ninth percentile in the Recreating Sentences subtest of the Test of Language Competence. This subtest requires a subject to expand language from two provided words into full sentences which are syntactically and semantically correct.

Description of Behavior

During Lynne's evaluation several behaviors were noted, including difficulty in word finding. The correct term was often produced after a process of circumlocution; a process which required a longer response time than is standardly tolerated. Oral expression seemed similar to hyper-verbalization. Lynne responded to relevant aspects of the environment, but she perseverated previous approaches to new stimuli.

Frequently, the materials or task success produced a response of excitement, requiring constant redirection. Lynne can be described as a child in constant motion, but she appeared oblivious to it. Although capable of maintaining work on a performance task, her physical position alternated between sitting, standing (on alternate feet), and climbing on the desk. She exhibited both gross motor activity and fine motor manipulation of anything available, activities which did not appear to distract the subject.

The evaluation presented above suggests that Lynne meets the criteria of a child with a specific learning disability, (New York State Education Department, 1980), based on average intelligence, perceptual processing deficits, academic underachievement and the absence of a primary emotional disturbance.

This case documentation may be considered interesting, and it could be dismissed as an isolated instance. That is, Lynne could be considered a unique case, except that available epidemiological data suggests that components of her birth defects are far from isolated, in that Lynne's subsequent physiological and psychological manifestations follow a pattern previously suggested by animal research. Research of environmental toxin exposure could be greatly enhanced and simplified by the emergence of a clinopathic syndrome that parallels experimental results. The outline of such a syndrome is found in this case study.

Environmental ambient exposure often occurs in environs where a mixture of toxic chemicals is found. Multiple exposure to these chemicals and identification of the most toxic concentration create complex problems. In this instance, the child was prenatally exposed to dioxin (TCDD) as well as other environmental toxins. Many forms of dioxins and related dibenzofurans exist, however, acutely toxic TCDD is considered the bellwether isomer with respect to research of health effects (Lowrance, 1983). In the remainder of this chapter I will discuss the epidemiological data from Love Canal as it is related to the experimental dioxin research. This discussion is limited to areas illustrative of Lynne's case; namely reproductive and teratogenic manifestations, weight and growth effects, kidney disorders, and central nervous system involvement such as spontaneous activity/epilepsy.

Teratogenic Defects

McConnell (1984) suggests that, in general, young animals, young humans, and females may be more susceptible to intoxication from dioxins than are adults and males. Several researchers have cited TCDD as one of the most potent teratogens (Murray, Smith, Nitschke, Humiston, Kociba, and Schwetz, 1979; McNulty, 1984; Mattison, Nightingale, and Silbergeld, 1983; Butcher and Page, 1981). The first indication of the teratogenic potential of TCDD was demonstrated in the study of mice and rats by Courtney and Moore (1971) in which simple 2,4,5-T was contrasted with combined 2,3,5-T with TCDD. Effects of cleft palate and cystic kidneys in offspring were shown to be related to the TCDD contaminated 2,4,5-T. Similarly, when Giavini (1982) studied the effects of dioxin administered to pregnant rats during the preimplantation period, he found an increase in the incidence of malformation and a decrease in body weight of fetuses from treated mothers which could be due to the storage of the toxicant in maternal organs or in the embryo itself. Giavini suggested that TCDD may manifest its effects a various stages of pregnancy.

Epidemiological data based on human populations are derived primarily from a small number of studies. In Seveso, Italy in 1979, for example, an ICMESA plant which produced hexachlorophene exploded. Later, Tenchini, Crimaudo, Pacchetti, Mottura, Agosti, and DeCarle (1983) reported detection of chromosomal abnormalities in aborted fetus tissues. Garattini (1982) indicated that children born to exposed women showed a higher than average incidence of anencephaly, spina bifida, hydrocephaly, polydactyly, hypospadia, and hemangiomas. However, as with many such studies with human subjects, the design calls the results into question. Reggiani (1980), on the other hand, reports no abnormalities in the somatic development of infants from this population.

LOVE CANAL DATA

A major study on birth defects in the Love Canal area was published in 1985 by Goldman, Paigen, Magnant and Highland. In this research, birth defects, malformations and deformations were defined according to the classification of Smith(1981). Minor birth defects, mental retardation, and cerebral palsy were excluded unless accompanied by major birth defects. The exposed group is considered those children whose parents spent any portion of their gestational life residing in the Love Canal area. Data was excluded for siblings outside the ages of 1.5 to 16.9 years, children of families who moved prior to June 1980, and children who died or stillborn fetuses. Table 2-1 lists the results of this research.

The adjusted odds ratio, a statistic derived from logistic regression parameters, relates the relative risk of the exposed group and the risk of the control group. Taking the first factor, the exposed homeowner group reported that low birth weight was found in 11.1 percent of their offspring, and in 4.8 percent of the control offspring. The odds ratio is adjusted and indicates approximately 3 times the risk for exposed homeowners to have low birth weight babies. For renters, this risk was a factor of one, or virtually no increased risk over controls for low birth weight babies.

In general, the population of Love Canal was characterized by two groups, the homeowners and the renters. Each group was distinctive in terms of its racial and socioeconomic characteristics, and they were matched with comparable groups in the same city. As a nonresident, Lynne would have been excluded from the study but, because her

Table 2-1.
Prevalence and Adjusted Odds Ratios of Low Birth Weight, Prematurity, Birth Defects, Malformations and Deformations by Residence in Love Canal and Control Homeowner and Renter Areas.*

	Love Canal	*(%)*	*Control*	*(%)*	*Adjusted Odds Ratio*	*95% Confidence Interval*
Homeowners						
Low Birth Weight	14/126	(11.1)	17/356	(4.8)	3.05	(1.32, 7.06)
Prematurity	24/130	(18.5)	53/351	(15.1)	1.41	(0.80, 3.48)
Birth Defects	20/134	(14.9)	31/354	(8.5)	1.95	(1.03, 3.72)
Malformations	13/134	(9.7)	18/354	(5.1)	2.42	(1.07, 5.47)
Deformations	8/134	(6.0)	13/354	(3.7)	2.68	(0.96, 7.07)
Renters						
Low Birth Weight	13/ 94	(13.8)	43/341	(12.6)	1.06	(0.50, 2.26)
Prematurity	16/ 96	(16.7)	54/353	(15.3)	1.11	(0.56, 2.20)
Birth Defects	9/105	(8.6)	13/354	(3.7)	2.87	(1.15, 7.18)
Malformations	4/105	(3.8)	11/354	(3.1)	1.96	(0.40, 8.30)
Deformations	5/105	(4.8)	2/354	(0.6)	8.67	(1.53,47.30)

*Number of missing values: birth weight, 34; prematurity, 17.

Source: Goldman, Paigen, Magnant, and Highland, 1985.

grandparents were homeowners, this population is relevant to our case study.

The data on birth defects shown in Table 2-1 refers to 239 children whose mothers spent a part of their gestational period in the Love Canal area, and 708 children whose gestational period occured elsewhere. Regarding low birth weight infants, Goldman et al. (1985) reported that there was no difference in the proportion of low birth weight babies from homes which were less than 240 meters from the Canal compared to those from homes further away. For the wet areas of the Canal, however, 10 out of 57 babies (17.5 percent) were of low birth weight compared to 8.6 percent for controls ($p < .04$). For homeowners located in areas classified as wet, 8 out of 50 babies (16%) were low birth weight compared to 4.8 percent of the controls ($p < .005$). The wet area homes were located on the natural drainage pathways in the area. The incidence of low birth weight was independent of other factors such as socioeconomic status, race, high maternal smoking, alcohol consumption, illness, or maternal age. The increased occurrence of low birth weight was accounted for almost entirely by women residing in wet area homes which were located close to the paths of chemical migration.

Research with TCDD in all nonhuman species shows lesions in the thymus (McConnell, 1984). Thyroid hormones have many regulating responsibilities within an organism. They monitor metabolic rate, development and growth, water and electrolyte mechanisms, lipid metabolism, CNS functions, skeletal muscles, and the cardiovascular system. Although involution of the thymus (Gupta, Vos, Moore, Sinkl and Bullock, 1973; Harris, Moore, Vos and Gupta, 1973) loss of body weight, and the appearance of the "wasting" syndrome at toxic doses (Seefield, Corbett, Kessey, and Peterson; 1984) are the most consistent symptoms, the mechanism is still unclear (Matsumura, Brewster, and Madhukar, 1983). Growth retardation, decreased survival, and smaller size at birth have been reported in animals subjected to TCDD (Moore, Gupta, Zenke, and Vos, 1973; Murray et al. 1979).

Vianna and Polan (1984) compared Love Canal birth weights with a more general New York state population. They report that the number of low birth weight infants born from the swale area was significantly higher than both upstate New York ($p < .009$) and the urban area ($p < .029$). Swale area residents had significantly more low birth weight children than did residents of the rest of the Canal ($p < .027$). However, this finding was for the period of active dumping, from 1940 to 1953, and the results differ from the Paigen study for the period in which no formal dumping occured (1965 to 1978). Although Vianna and Polan (1984) found no greater risk was involved for the later period, Goldman et al. (1985) suggest that the use of a generic control group influenced the calculation.

Paigen, Magnant, Goldman, Highland, and Steegman (1987) tried to determine whether living near a hazardous waste site had an affect on growth. In this study two groups were compared: 493 Love Canal children and 428 matched controls, who were similar on anthropomorphic measures, health information, and stature of parents.

Age and weight percentiles were measured by the Stature Age Percentile Ranking (SAPR) and Weight Age Percentile Ranking (WAPR) standard scales. For prepubertal children, the two factors associated with significantly smaller stature were birth in the area and fraction of life exposed to the Love Canal neighborhood. Of the groups studies, those born and spending 75 percent of their lives or more in the Love Canal area were most adversely affected. Exposed children had a mean stature for age percentile of 46.4 ± SE 2.2 compared to 53.0 ± SE 1.0 for controls ($p < .004$). This difference could not be associated with midparent height, socioeconomic class, nutrition, birth weight, or chronic illness. The

Table 2-2.
Mean sex and age specific percentiles for stature [SAPR], weight [WAPR], and weight-for stature [WSPR] by different measures of exposure.

Exposure Classification		Stature	Weight	Weight-for-stature
		n Mean + S.E.	*n Mean + S.E.*	*n Mean + S.E.*
I. Live in Control Area		428 53.1 ± 1.4	428 53.1 ± 1.4	238 52.4 ± 1.8
II. Live in Love Canal Area		493 50.8 ± 1.3	489 53.8 ± 1.3	195 54.5 ± 1.8
	p-value a	0.10	n.s.	n.s.
III. Born in Control Area		404 53.3 ± 1.4	404 53.2 ± 1.4	221 52.4 ± 1.8
IV. Born in Canal Area		213 48.3 ± 2.0	211 51.2 ± 2.0	121 54.5 ± 2.3
	p-value b	0.02	n.s.	n.s.
V. Born an 75% of life in Love Canal		172 46.6 ± 2.2	170 50.1 ± 2.2	99 54.8 ± 2.4
	p-value c	0.004	0.08	n.s.

a. There are 14 children with no information regarding birth.

b. Comparing groups III and IV.

c. Comparing groups III and V.

Source: Paigen, Magnant, Goldman, Highland, and Steegman, 1987.

weight for age percentile was 50.1 ± SE 2.2 as compared to 54.0 ± SE 1.0 for control ($p < .05$).

The response by the renal organ to toxin exposure has been reviewed by Finn (1977), and demonstrates a lack of research of the environmental toxins, including TCDD. Teratogenic research conducted by Neal, Gasiewicz, Geiger, Olson, and Sawahata (1984) includes renal malformations, a topic not widely researched. However, incontinence was among the list of seven health problems surveyed in the Love Canal area (Paigen et al., 1985) and is a symptom for Lynne. In addition, Lynne was diagnosed as epileptic and hyperkinetic, two of the seven major health problems found Paigen et al. (1985) to be significantly prevalent in the Love Canal area.

Learning Problems

Research on the learning problems experienced among children of Love Canal is inhibited by the generality of the definition used for a learning problem. Paigen et al. (1985) defined learning problems as existing in a child in a special class, receiving remedial assistance or repeating a grade. Using this definition, the high level of 28 percent incidence in Love Canal children versus 19 percent of the control children is not reflective of the

Table 2-3.
Adjusted Odds Ratios for Health Problems Prevalent in Love Canal Area

	Exposed		*Controls*		*Adjusted Odds Ratio*[1]	
	n	*positive*	*n*	*positive*	*95% percent C.I.*	
Seizures	520	5.2	440	1.8	2.45	(1.05– 5.75)
Learning Problems[2]	422	28.1	301	19.3	1.51	(1.04– 2.19)
Hyperactivity[3]	238	9.7	259	3.9	2.95	(1.3– 6.67)
Eye irritation	520	12.7	440	6.4	1.96	(1.20– 3.22)
Skin rashes	520	27.3	440	18.4	2.25	(1.43– 3.53)
Abdominal Pain[4]	487	11.7	391	6.1	2.09	(1.25– 3.51)
Incontinence[4]	487	3.1	391	1.0	3.27	(1.02–10.51)

1. Adjusted odds ratio and 95% confidence interval [C.I.] are derived from logitic regression parameters. Small differences from the total n of 963 are due to responses of "do not know."

2. Includes only children > 60 months

3. Includes only homeowner children.

4. Includes only children > 36 months.

Source: Paigen, Golden, Highland, Magnant, & Steegman, 1985.

federal guidelines of a learning disability; yet Lynne's behavior complies with the more stringent federal definition. Still, these data do suggest that exposure to environmental contaminants in a chronic, low-level nature may have some influence on the learning progress of Love Canal children, and that this issue is deserving of further exploration.

SUMMARY AND CONCLUSIONS

The purpose of this chapter is to illustrate the paucity of research, particularly of the subtle and insidious effects of toxins on learning skills and behavior of children. Scientists skilled in methods of analysis and risk assessment should not ignore these situations.

In this report, the results of laboratory studies involving TCDD exposure have been compared to studies of children at Love Canal and as expressed in our case study subject, Lynne. Love Canal was not a laboratory situation; thus exposure to multiple chemical is a confounding factor. However, most humans do not exist in an experimentally clean situation prior to or during a toxin exposure.

The case of Lynne illustrates several somatic effects of toxin exposure; low birth weight, slow growth, birth defects, and renal disruption. Some of these have been researched as specific effects of TCDD exposure. Others are known to be areas of toxin influence, but research specific to

TCDD seems to be lacking. Nonetheless, this child, raised at a distance from the Love Canal area, shows a cluster of many major health problems similar to those identified in children of exposed homeowners.

The cluster includes several behavioral and cognitive difficulties which inhibit the development of her otherwise normal abilities. Several of these difficulties, broadly defined, were also present in the Love Canal population. Due to the presence of severe somatic birth defects, several alternative explanations for Lynne's behavioral and cognitive deficits are readily available. However, if none of the organ deficits were known, Lynne would present as an individual with moderate to severe cognitive and behavioral disorders of the central nervous system. It seems reasonable to argue that we must guard against a Type II error in our consideration, and be willing to consider the possibility that teratogenic agents may also produce CNS-related effects in behavior and cognition.

Further research is required in this area, and the study by Needleman (1979) of children exposed to lead could serve as a model for research on exposure to hazardous toxic waste. Research in the area of subtle behavioral effects of human toxin exposure is recent, and procedures which are both effective and efficient have yet to be standardized.

REFERENCES

Beatty, L.S., Madden, R., and Gardner, E.F.: *Stanford Diagnostic Arithmetic Test.* New York, Harcourt, Brace, Jovanovich, 1976.

Butcher, R. L., and Page, R. D.: Introductory remarks, environmental and endogenous hazards to the female reproductive system. *Environmental Health Perspectives, 38:*35, 1981.

Courtney, K. D., and Moore, J. A.: Teratology studies in 2,4,5 and 2,3,7,8 TCDD. *Toxicology and Applied Pharmacology, 20:*396, 1971.

Department of Health, Education and Welfare, US Committee to Coordinate Environmental and Related Programs: *Report of the Subcommittee on the Potential Health Effects of Toxic Chemical Dumps,* Appendix D, Chemicals Found in Waste Dumps (Love Canal). 1980.

Dunn, L.M., and Dunn, L.M.: *Peabody Picture Vocabulary Test-Revised.* Circle Pines, MN, American Guidance Service, 1981.

Environmental Protection Agency: *Environmental Monitoring at Love Canal.* Vol. 1 and 2, Washington, D.C. 1982.

Finn, W. F.: Renal Response to Environmental Toxins. *Environmental Health Perspectives, 20:*15–26, 1977.

Garattini, S.: TCDD toxicology with particular reference to Seveso. *Drug Metabolism Reviews, 13*: 345-469, 1982.

Gardner, M.: *Expressive One-Word Picture Vocabulary Test.* Novato, CA, Academic Therapy Publications, 1982.

Giavini, E., Prati, M., and Vismara, C.: Effects of 2,3,7,8 tetracholorodibenzo-p-dioxin administered to pregnant rats during the preimplantation period. *Environmental Research, 29:*185–189, 1982.

Goldman, L. R., Paigen, B., Magnant, M., and Highland, J. H.: Low birth weight, prematurity, and birth defects in children living near the hazardous waste site, Love Canal. *Hazardous Wastes and Hazardous Materials, 2:*209–223, 1985.

Gupta, A. B., Vos, J. C., Moore, J. A., Zinkl, T. G., and Bullock, B. C.: Pathological effects of 2,3,7,8 tetrachlorodibenzo-p-dioxin in laboratory animals. *Environmental Health Perspectives, 5:*125–140, 1973.

Harris, M. W., Moore, J. A., Vos, J. C., and Gupta, A. B.: General biological effects of TCDD in laboratory animals. *Environmental Health Perspectives, 5:*101–109, 1973.

Haymanker, W.: Morphological changes in the nervous system following exposure to ionising radiation. In Weis, B., and Laties, G. (Eds.): *Behavioral Toxicology.* New York, Plenum, 1975.

Hicks, S. P., D'Amato, C. J., and Lowe, M. L.: The development of the mammalian nervous system. *Journal of Comparative Neurology, 113:*435–469, 1959.

Karlson, B., Madden, P., and Gardner, E.F.: *Stanford Diagnostic Reading Test.* New York, NY, Harcourt, Brace, Jovanovich, 1976.

Kirk, S., McCarthy, J., and Kirk, W.D.: *Illinois Test of Psycholinguistic Abilities.* Urbana, IL, University of Illinois Press, 1968.

Koppitz, E. M.: *Bender Visual Motor Gestalt Test for Young Children.* New York, NY, Grune and Stratton, 1964.

Lenz, W.: Thalidomide and congenital abnormalities. *Lancet, 1:*45, 1962.

Leonard, R. P., Werhmary, P. H., and Zieglon, R. C.: Characterization and abatement of ground water pollution from Love Canal chemical landfill, Niagara Falls, New York. *Carlspan Report #N D 6097-M-1.* Buffalo, NY, Carlspan Company, 1977.

Lilis, R.: Warning on Solvents in Consumer Products. *Nutrition Health Review, 43:*7, 1987.

Lowrance, W. W. (Ed.): *Health Risks of the Dioxins.* New York, Rockerfeller University Press, 1984.

McConnell, E. E.: Clinopathologic concepts of dibenzo-p-dioxin intoxication. *The Banbury Report.* Cold Spring Harbor Laboratory, 1984, pp. 27–38.

McNulty, W. P.: Fetocidal and teratogenic actions of TCDD. In Lowrance, W. W. (Ed.): *Public Health Risks of the Dioxins.* New York, Rockefeller University, 1984, pp. 245–254.

Matsumura, F., Brewster, D., and Madhukar, B. V.: Alteration of fat hepatic plasma membrane functions by TCDD. In Lowrance, W. W. (Ed.): *Public Health Risks of the Dioxins.* New York, Rockefeller University, 1984, pp. 295–314.

Mattison, D. R., Nightingale, M. S., and Silbergeld, E. K.: Reproductive toxicity of tetrachlorodibenzo-p-dioxin. In Lowrance, W. W. (Ed.): *Public Health Risks of the Dioxins.* New York, Rockefeller University, 1983. pp. 217–244.

Moore, J. A., Gupta, B. N., Zenke, T. G., and Vos, T. C.: Postnatal effects of maternal exposure to TCDD. *Environmental Health Perspectives,* 5:81–85, 1973.

Murray, F. J., Smith, F. A., Nitschke, K. D., Humiston, C. G., Kociba, R. T., and Schwetz, B. A.: Three generation reproduction study of rats given 2,3,7,8 tetrachlorodibenzo-p-dioxin (TCDD) in the diet. *Toxicology and Applied Pharmacology, 50:*241–252, 1979.

Nair, V., and Dobois, K. P.: Prenatal and early postnatal exposure to environmental contaminates. *Chicago Medical School Quarterly, 27:*75, 1968.

Neal, R. A., Gasiewicz, T. A., Geiger, L., Olson, J. and Sawahata, T.: Metabolism of 2,3,7,8-tetrachlorodibenzo-p-dioxin in mammlaian systems. In Poland and Kimbrough (Eds.): *The Banbury Report.* Cold Spring Harbor Laboratory, 1984, pp. 204–213.

Needleman, H. L., Gunnoe, C., Leviton, A., Reed, R., Peresie, H., Maher, C., and Barrett, P.: Deficits in psychologic and classroom performance of children with elevated dentine lead levels. *New England Journal of Medicine, 300:*689–695, 1979.

New York State Education Department: *Serving the L D Child.* Albany, NY, University of the State of New York, 1980.

Paigen, B.: Health hazards at Love Canal. Testimony presented to the House Subcommittee on Oversights and Investigations. Unpublished manuscript, 1979.

Paigen, B., Goldman, L. R., Highland, T., Magnant, M. M., and Steegman, A. T.: Prevalence of health problems in children living near Love Canal. *Hazardous Wastes and Hazardous Materials, 2:*23–43, 1985.

Paigen, B., Magnant, M. M. Goldman, L. R., Highland, T. H., and Steegman, A. T.: Growth of children living near hazardous waste site, Love Canal. *Human Biology,* 59:489–508, 1987.

Reggiani, G.: Localized contamination with TCDD—Seveso, Missouri, and other areas. In Kimbrough, R. D. (Ed.): *Halogenated Biphenyls, Teraphenyls, Nephaltralenes, Dibenzo-dioxins and Related Products.* Amsterdam, Elsevier, 1980, pp. 303–372.

Rugh R., and Wohlfromm, M.: Previous reproductive history and the susceptibility to x-ray induced congenital abnormalities. *Nature, 210:*969–970, 1966.

Seefield, M. D., Kessey, R. E., and Peterson, R. E.: Body weight regulation in rats treated with 2,3,7,8 tetrachlorodibenzo-p-dioxin. *Toxicology and Applied Pharmacology, 73:*311–322, 1984.

Smith, D. W.: *Recognizable Patterns of Human Deformation.* Philadelphia, PA, W. B. Saunders, 1981.

Spyker, J. M.: Behavioral Teratology and Toxicology. In Weis, B., and Laties, V. G. (Eds.): *Behavioral Toxicology.* New York, Plenum, 1975.

Tenchini, M. L., Crimaudo, C., Pacchetti, G., Mottura, A., Agosti, S., and DeCarle, L.: A comparative cytogenetic study on cases of induced abortions in TCDD exposed and non-exposed women. *Environmental Mutogenesis, 5:*73–85, 1983.

Terman, L. M. and Merrill, M. A.: *Stanford-Binet Intelligence Scale.* Boston, Houghton Mifflin, 1960.

Vianna, N. J., and Polan, A. K.: Incidence of low birth weight amoung Love Canal residents. *Science, 226:*1217–1219, 1984.

Wechsler, D.: *The Wechsler Intelligence Scale for Children—Revised.* New York, Psychological Corporation, 1974.

APPENDIX

A. Partial Listing of Chemicals Found at Love Canal

Benzene
Carbon Tetrachloride
Vinyl Cloride
Dichloroethanes
Hexachlorobenzene
Hexachlorocyclohexanes (6350 metric tones of the λ-isomer, lindane, known for effects of hyperirratibility and convulsions)
Polychlorinated Biphenyls
Trichlorophenols (180 metric tons)
Tetrachlorodibenzo-p-dioxin (TCDD, 2,3,7,8 configuration)
Toluene
Xylenes

B. Review of Known Toxicities of the 248 Substances Found

36 Neurotoxins
4 Pulmonary toxins
20 Hepatotoxins
15 Renal toxins
34 Carcinogens
18 Teratogens
30 Fetotoxins or embryotoxins
157 Identified toxins of 248 waste substances

Source: Paigen, Goldman, Highland, Magnant and Steegman, 1985

Chapter 3

PERCEPTION OF RISK AND TOXIC HAZARDS

WILLIAM HALLMAN AND ABRAHAM WANDERSMAN

INTRODUCTION

Nearly every day one encounters news reports detailing another threat to public health and well-being. We are constantly bombarded with frightening stories about hazardous chemicals, nuclear wastes, cancer causing pesticides, radon gas, holes in the ozone layer, leaking tankers, and a thousand other threats to the air we breathe, the water we drink, the food we eat, and the place we live. There are often reports about newly discovered links between cancer and one of the many chemicals designed to enhance our daily lives. Perhaps just as often, however, we read stories relating government or industry officials' views that particular risks are "inconsequential," or "nothing to worry about," or "almost negligible." Because of these conflicting news reports, it is often difficult to make conclusions about what risks people really need to worry about.

The sources of information surrounding hazardous waste disposal are particularly confusing. As a result, it is not surprising to find a diversity of opinions concerning the risks posed by hazardous waste facilities. Risk perception is more than estimating probabilities of harm, it is a complex, affect-laden process which can be highly individualistic. Close friends and neighbors often have very different beliefs concerning the risks posed by such facilities. Similar disagreements occur among facility operators, government regulators, health officials, and community action groups.

The purpose of this chapter is to provide information which helps to explain this diversity in perceptions of risk. After a brief discussion of the importance of perceived risk, the first section of the chapter reviews conceptual approaches to the perception of risk. The second section explores the empirical literature on community and individual differences which influence residents' perceptions of risk. Finally, we summarize our observations and suggest directions for future research.

THE IMPORTANCE OF PERCEIVED RISK

The fact that perceptions of risk often have important consequences makes understanding what influences these perceptions especially important. Issues of perceived risk are especially significant to public policy makers. The perceived acceptability of a risk is systematically correlated with public actions supporting or opposing new or existing technologies (Gardner, Tiemann, Gould, DeLuca, Doob, and Stolwijk, 1982). In addition, much of the Not In My Back Yard (NIMBY) Syndrome is directly related to perceived risk (Edelstein, 1988). As a result, effectively shaping private and public perceptions of risk can determine whether new technologies, processes, or facilities are put into place (Elliott, 1984). Perceptions of risk can also influence the adoption of policies which might mitigate the effects of environmental disaster (Mileti, 1980). Risk perceptions can also have other economic impacts. Wide-spread perceptions of risk can stigmatize whole communities, lowering property values and influencing newcomers' decisions to settle in a community and long-time residents' decisions to remain.

Understanding perception of risk is important from philosophical, moral, and ethical standpoints. Perceptions of risk implicitly involve personal and societal value judgements. In order to define how risky something is, one must identify its adverse effects on those things which are considered valuable. Fischhoff, Watson and Hope (1984) argue that defining risk should not be left solely to scientists, since science provides no special insights into what society should value. Moreover, scientific definitions of risk often entail the use of untested assumptions. Sjoberg (1980) warns that even quantitative risk analysis is enveloped in uncertainty. As a result, Fischhoff et al. (1984) suggest that there is no definition of risk which is suitable for all problems. They suggest that definitions of risk must be generated which are relevant to specific circumstances. These definitions must take into account public perceptions of risk.

Perceived risk can also have psychological impacts. Perceptions of risk associated with toxic hazards have been correlated with psychological distress, fear, and anxiety (Hallman, in progress). An understanding of what influences perceived risk can help professionals deal with these negative outcomes in communities affected by environmental hazards.

CONCEPTUALIZATIONS OF RISK

One approach to comprehending perceptions of risk has been to create taxonomies of hazards. These classification schemes are used to compare, understand, and predict responses to particular risks. In addition, they may be used to explain why some risks are overestimated and others are ignored. Several methods have been employed to create these hazard classifications.

Natural Vs. Man-Made Disasters

In his book *An Anatomy of Risk,* Rowe (1977) describes four types of hazards that could be distinguished according to the role humans played in their cause or prevention. Some hazards such as erupting volcanos are completely beyond the dominion of human beings and are called "unavoidable natural disasters." "Natural disasters, as agents of psychosocial stress, are euphemistically referred to as 'acts of God,' a phrase which essentially eliminates human agency or culpability from analysis" (Bolin, 1987, p. 6). Other natural disasters can be avoided or their effects lessened by proper human precaution or action. These are referred to as "avoidable natural disasters," and would include hazards such as floods or mudslides.

Rowe divides man-made hazards or "man-originated disasters" into two classes. One class includes catastrophes such as the Buffalo Creek Disaster (Erikson, 1976). In Buffalo Creek, men were responsible for creating an inadequate dam but naturally occurring high water ultimately caused the dam to fail. Rowe classified such disasters as "man-triggered disasters." Conversely, "man-caused disasters" are those in which man is responsible for both creating the circumstances that make the disaster possible and engaging in activities which directly causes the disaster. The release of radioactive gas at Three Mile Island might be categorized as "man-caused."

Baum, Fleming, and Davidson (1983) classified hazards as "natural disasters or technological disasters" according to five characteristics. They argued that natural disasters typically create visible property damage; recur with some predictability; have a distinguishable low point at which "the worst is over"; are not seen as controllable; and have effects which are usually short lived and are limited to the immediate victims of the disaster. In contrast, technological disasters do not always create visible property damage, particularly if they involve radiation or chemicals.

Technological disasters are usually not predictable and do not have clear low points. In fact, conditions may not necessarily improve over time. As opposed to natural disasters where people are reminded of their *lack* of control over the elements, technological disasters are likely to be perceived as the *loss* of control. Moreover, technological disasters may produce both short-term or long-term effects, but are most likely to cause chronic difficulties for its victims. Such disasters can also have effects on people not directly victimized by the event, such as promoting a loss of confidence in government and science.

Couch and Kroll-Smith (1985) suggest that in contrast to natural disasters which have immediate impacts "concentrated in time and space," chronic technical disasters are "a slowly developing, extended, humanly-produced deterioration in human system-ecosystem relations . . . (p. 566)." They also suggest that because chronic technical disasters emerge slowly, they tend to create a great deal of uncertainty. As a result, disagreements often arise about how to interpret what is occurring and what to do about it.

Miller and Fowlkes (1984) argue that the term "technological disaster" is a misnomer which fails to accurately characterize the causal role played by humans. They suggest that the term "man-made disaster" be substituted. In addition, all disasters do not fit neatly into the natural/technological classification system. Cvetkovich and Earle (1985) contend that the lack of agreement concerning which disasters may be "natural" and which should be considered "technological" makes this distinction less than utilitarian.

Comparative Risks

Taxonomic systems of comparative risks typically enumerate hazards and compare them according to their potential for reducing life expectancy. Reduction in life expectancy is useful for measuring and comparing risks because it takes into account the higher costs associated with the death of a young person. In addition, it permits the comparison of instant accidental deaths with the delayed health effects of hazards such as radiation (Reissland and Harries, 1979).

Cohen and Lee (1979) created such a taxonomy. They examined risks such as radiation, accidents, disease, being overweight, and the use of tobacco, alcohol, drugs, coffee, saccharin, and birth control pills. In addition, they looked at occupational risks, socioeconomic factors, marital status, geography, military service in Vietnam, catastrophic events, energy production, and technology. By expressing risks in terms of the number of days of life expectancy lost and ranking them according to their severity, they hoped to make risk more understandable to the public.

Another example of a taxonomy of comparative risks is a ranking of hazards conducted by extrapolating animal bioassays to humans. Ames, Magaw and Gold (1987) ranked carcinogens according to their possible health effects, based on the amount of known rodent carcinogens to which humans might be chronically exposed. The resulting scale was not proposed as providing an empirical basis for assessing risks to humans, but rather as a tool for priority setting.

Cost-Benefit Approaches

Starr (1969) studied hazardous activities which the public has historically accepted and examined the relationship between the risks and benefits derived from each. Starr assumed that over time, an optimum balance between the risks and benefits has evolved. In addition, these "revealed social preferences" are consistent enough to permit their use for predictive purposes. Using national accident records, Starr defined risk as the expected number of fatalities which resulted from exposure to one hour of the activity. He measured benefit by estimating the average sum of money spent on that activity by an individual participant, or the average increase the activity made to a person's income. The results of these comparisons yielded several observations concerning people's willingness to accept risk, that is, their revealed preferences.

He suggests that people are generally willing to accept risks which are roughly proportional to the amount of benefits derived raised to the third power. Given the same amount of benefit, Starr also found that risky activities which are voluntary, are about 1000 times more acceptable than those that are involuntary. In addition, awareness of the benefits of an activity directly influences the social acceptance of its risk. Finally, Starr suggests that the amount of risk considered acceptable for voluntary activities is about the same as the risk of death attributable to disease. He suggests that the statistical mortality due to involuntary exposure to disease may serve to define whether a risk taker is bold or merely foolhardy.

A number of authors have disagreed with Starr's conclusions. Fischhoff, Slovic, Lichtenstein, Read, and Combs (1978) contend that although

Starr's revealed preference method has the benefit of examining behaviors rather than attitudes, it assumes that past behavior is a valid measure of present preferences. Second, as Starr himself argues, the revealed preference approach only discloses what has been traditionally acceptable in terms of the market place. It does not necessarily reflect what is "best" for society or the public's preferences regarding safety. Third, the revealed preference approach presupposes that the public has full information concerning the risks associated with the activities. However, as Torry (1979) points out, there are many segments of the population who have little experience or understanding of hazards. Even when individuals are fully aware of a hazard, there is no guarantee that people are able to use that information or that they will act in accordance (Sims and Baumann, 1983).

Philipson (1986) catalogued other criticisms of Starr's findings. First, it is possible that many of the risk-takers did not understand the risks they were taking. As a result, the fact that risky activities were accepted in the past does not argue for accepting them in the present. Second, many of the risk-takers who were thought to engage in activities "voluntarily," may have had few alternatives. Philipson suggests that over time, new alternatives may evolve and that the risks posed by a given activity may change. Third, Starr's method of using average risk and average benefit obscures the possibility that those who are accepting the greatest risks may not be those receiving the greatest benefit. As a result, Starr fails to consider the disproportionate impacts of specific risks.

Psychometric Approaches

Another approach to creating taxonomies of hazards is the psychometric paradigm. In contrast to Starr's revealed preferences method, the psychometric paradigm uses multivariate analyses of psychophysical ratings to create quantitative "cognitive maps" of risk attitudes and perceptions (Slovic, Fischhoff and Lichtenstein, 1986).

Fischhoff et al. (1978) used this "expressed preferences," psychometric approach to evaluate the perceived risks and benefits of risky activities. Subjects rated the particular risks and benefits to society associated with each of thirty different activities or technologies. Subjects also rated the acceptability of the given technology or activity given its current level of risk and its standing on each of nine dimensions of risk. These dimensions were: (1) the voluntariness of the risk; (2) the immediacy of the effects; (3) the extent to which the risks are known to those exposed; (4) the extent to which the risks are known to science; (5) the amount of personal control one has over avoiding death while engaging in the activity; (6) the newness of the threat; (7) the extent to which the risk is chronic or catastrophic; (8) the extent to which the risk is common or dreadful; and (9) the severity of the consequences of the risk.

Principal components factor analysis revealed two orthogonal factors which seemed to account for the shared variances among the nine dimensions measured. The first factor, called "technological risk," discriminated between high and low technology activities. High technology activities tend to be new, poorly known, involuntary, and often have delayed consequences. Low technology activities tend to be older, more familiar, voluntary, and have more immediate consequences. The second factor, named "severity," divided those activities that would result in certain fatalities should an accident happen, from those with fewer dire consequences.

As a result of their analyses, Fischhoff and his colleagues found that perceived risk correlated with the severity of the risk and the extent to which it was viewed with dread. Perceived risk did not correlate with voluntariness, as Starr (1969) predicted, nor any of the other six dimensions measured. Voluntariness did predict the *acceptability* of risk, however. In addition, they found that given the same amount of benefit, greater risks were considered acceptable if they were voluntary, immediate, well known, controllable, and familiar.

Slovic, Fischhoff, and Lichtenstein (1980) used the same methods, but with a list of ninety activities. They uncovered a third factor which discriminated activities according to the number of people affected by the risks of the activity.

Johnson and Tversky (1984) used ratings of similarity among eighteen hazards to compare cognitive representations of risks derived from factor analytic methods, multidimensional scaling methods, and cluster analysis. They found that their factor analysis model coincided with those obtained in other other factor analytic studies. However, the model derived from the multi-dimensional scaling analysis differed from the factor analytic solution. In the multi-dimensional scaling representation, two hazards judged to be quite similar differed on the risk dimensions identified by the factor analysis.

Cluster analysis revealed yet another representation of the similarities among hazards. It yielded a structure that Johnson and Tversky labeled:

diseases (such as stroke or cancer); technological disasters (such as a nuclear accident); violent acts (such as war or homicide); accidents (such as accidental falls or airplane accidents); and hazards (such as fire or electrocution).

Hohenemser, Kates, and Slovic (1983) also attempted to identify the distinguishing characteristics of technological hazards. Ninety-three hazards were scored on twelve descriptive scales. These were:

1. Intentionality, the degree to which a technological activity is intended to cause harm;
2. Spatial extent, the maximum distance over which hazard has significant impact;
3. Concentration, the amount of material or energy released compared to natural background levels;
4. Persistence, the amount of time a hazardous release is a significant threat to humans;
5. Recurrence, the mean time interval between releases;
6. Population at risk, the number of people in the U.S. potentially at risk due to the hazard;
7. Delay, the amount of time between a hazardous release and the onset of consequences;
8. Human mortality (annual), the average annual deaths in the U.S. due to the hazard;
9. Human mortality (maximum), the maximum number of deaths credited to a single hazardous event;
10. Transgenerational, the number of future generations considered at risk from the hazard;
11. Nonhuman mortality (potential), the maximum potential mortality to nonhumans;
12. Nonhuman mortality (experienced), the nonhuman mortality actually experienced.

The authors characterize technological hazards as a sequence of causally connected events. These events start with human wants and needs and the selection of a technology to meet them. At this stage, intentionality differentiates hazards. Next in the causal sequence is the release of materials or energy. At this step, hazards are distinguished by spatial extent, concentration, persistence, and recurrence. The next step in the sequence is exposure to the materials or energy. Here, hazards are differentiated by the population at risk and the delay of consequences.

Finally, after exposure, there are human and biological consequences. At this stage, hazards can be distinguished by annual and maximum human mortality, transgenerational impact and potential, and experienced nonhuman mortality.

Principal components factor analysis reduced the original twelve scales to five factors. The first factor, biocidal, incorporates potential and experienced nonhuman mortality and intentionality. The second factor was termed delay. It combines the factors of persistence, delay, and transgenerational effects. The third factor, catastrophic, accounts for the recurrence and human mortality (maximum) descriptors, while the fourth factor, mortality, considers human mortality (annual). The final factor, global, combines the population at risk with concentration descriptors.

To test whether the twelve descriptors and the five factors derived from them could predict perceptions of risk, subjects rated the risk of dying across all of U.S. society as a result of the hazard in question. Modest positive correlations were calculated between perceived risk and nine of the twelve descriptors and four of the five factors. Risk perception did not correlate with the concentration, recurrence, or human mortality (annual) descriptors, nor the mortality factor.

The psychometric means of studying risk perception has been criticized. Covello (1983) reports that most of the risk perception findings have been based on surveys of small, highly specialized, and unrepresentative groups. Douglas and Wildavsky (1982) argue that because it is primarily a cognitive approach, the psychometric method does not consider the social processes which also influence the acceptability of specific risks.

Perceptions of risk also seem to be quite sensitive to the way in which information is presented (Tversky and Kahneman, 1981) and by the way risk judgements are collected. Earle and Lindell (1984), using a free response format to solicit public perceptions of industrial risks, obtained results different from those using more traditional closed question methods.

Cvetkovich and Earle (1985) identified several weaknesses in the psychometric approach that focus on the way risk judgements are measured. First, they argue that this approach often involves asking subjects to make comparisons between hazards that may not be meaningful, such as comparing the risks of pesticides and skiing. Second, psychometric tasks often include judging risks independent of their associated benefits or other contextually significant information. Third, little attention is given to the positive aspects of taking risks. Taking risks can be considered recreational as in sky-diving, bull-fighting, and other activities.

Although the psychometric approach has been criticized, this method demonstrates that risk perceptions can be quantified. As a result, psychometrics may be useful for comparing differences in risk perceptions between groups (Slovic, Fischhoff, and Lichtenstein, 1979). One of the most meaningful comparisons of perception of risk are those between the general public and experts. Hohenemser, Kates and Slovic (1983) and Slovic, Fischhoff, and Lichtenstein (1979) report that experts' perceptions of risk tend to focus on statistical measures of expected mortality rather than the dimensions that have been found to be salient in the general population. Moreover, although the fatality estimates made by the general population are fairly accurate, risk judgements made by lay people show only a moderate relationship to annual fatalities caused by risky activities. This suggests that one of the difficulties that experts and the public have in communicating may be based on different definitions of risk.

Other Approaches

Other theoretical conceptions of perception of risk exist based largely on reviews of the literature. Paterson and Neufeld (1987) identify several factors which may act as situational determinants of the primary appraisal of threat. Primary appraisal of the situation and assessment of the risks involved also have been cited as the crucial variable in the coping process (Lazarus, 1966; Lazarus and Launier, 1978; Lazarus and Folkman, 1984). As a result, understanding what influences these appraisals is important.

Paterson and Neufeld (1987) suggest that a person's primary appraisal of threat is influenced by: the severity of the threatened event; the importance of the goals threatened; the number of goals threatened; the extent to which goals are threatened; the imminence of the threat; the length of anticipation of the threat; and the probability of the threatening event. Based on their review of the pertinent literature, Paterson and Neufeld advanced several propositions regarding the relationship between these factors and the resulting anticipatory stress.

Proposition 1.0: The greater the severity of the threatened event, the greater the anticipatory stress.
Proposition 1.1: The more important a goal, the greater the anticipatory stress when it is threatened.
Proposition 1.2: The more goals that are threatened, the greater the anticipatory stress that is produced.

Proposition 1.3: The greater the extent to which a goal is threatened, the greater the anticipatory stress produced.

Proposition 2.0: Threat is proportional to severity, and Severity is equal to the summation of (goal importance$_i$ X intensity of deprivation$_i$).

Proposition 3.0: The more imminent stressor occurrence is, the greater its capacity to arouse anticipatory stress.

Proposition 4.0: The longer the anticipation time given, the greater the anticipatory stress when the event is imminent.

Proposition 5.0: The lower the probability of a specified event's occurrence, the lower the level of anticipatory stress it produces.

Paterson and Neufeld also hypothesize that severity, imminence, and probability are all necessary and together are sufficient to produce anticipatory stress. Also, incubation can increase stress levels but is not necessary for stress to occur.

Mileti (1980) analyzed much of the sociological literature dealing with hazards. He proposes that perceived risk is influenced by people's ability to estimate risk; the perceived cause of the disaster (God, nature, technology); past experience with risk; propensity to deny risk; access to information; and unit of analysis (a devastating flood will certainly happen each year in the U.S. but not in my community).

Generating perhaps the most extensive model of risk perception and acceptability, Vlek and Stallen (1980) suggest thirty-two specific "aspects of risk" that determine the nature and/or context of a risky decision problem. These are grouped under eleven categories. It is proposed that the acceptability of risky activity varies with: voluntariness of exposure; controllability of the consequences; distribution of consequences in time; distribution of consequences in space; context of probability assessment; context of accident evaluation; combination of accident-probability and "seriousness"; knowledge about the risky activity; condition of the subject; social considerations; and confidence in experts/regulators (See Table 3-1).

Each model is useful to integrate a diverse array of variables which have been suggested either empirically or theoretically to influence perceptions of risk. As a result, their contribution has been to create a framework for understanding perceptions of risk. Unfortunately, these models have gone largely untested.

Table 3-1.
Aspects of Risk

It is **hypothesized,** that the risks of an activity are (unthinkingly, tacitly, voluntarily, or perforce) more readily accepted to the extent that:

I. **Voluntariness of exposure**
 A. there are fewer comparable options available;
 B. someone's personal influence on the decision is larger;
 C. correction or change of the selected action is easier;
 D. the importance of intended benefits is larger;

II. **Controllability of consequences**
 A. accident probabilities can be influenced by safety measures;
 B. there is less room for human error, recklessness, and sabotage;
 C. accident-effects can be contained by rescue operations;
 D. accident-effects are reversible;

III. **Distribution of consequences in time**
 A. intended benefits are obtained at a shorter term;
 B. undesired consequences have delayed effects;

IV. **Distribution of consequences in space**
 A. intended benefits materialize at a shorter (social) distance;
 B. undesired consequences occur at a greater (social) distance;

V. **Context of probability assessment**
 A. judged (relative) accident-frequency is lower;
 B. imaginability of accident-causation is less;
 C. final assessment of accident-seriousness is less ambiguous;

VI. **Context of accident evaluation**
 A. an accident violates less central values or less basic needs;
 B. imaginability of final accident-effects is less;
 C. final assessment of accident-seriousness is less ambiguous;

VII. **Combination of accident-probability and 'seriousness'**
 A. risks are of the type 'relatively high-probability/small-loss;'

VIII. **Knowledge about the risky activity**
 A. one is better informed about intended benefits;
 B. one has heard, read, or seen less about the risks;
 C. one has a lesser relevant personal accident experience;

IX. **Condition of the subject**
 A. someone's permanent condition favors risk acceptance;
 B. someone's temporary condition favors risk acceptance;
 C. someone's accumulated 'risk-load' is lower (or: more optimal);

X. **Social considerations**
 A. the benefit/loss ration is relatively favorable to the subject;
 B. voluntariness of exposure is equitably distributed;
 C. controllability of consequences is equitably distributed;
 D. social traditions, norms and customs favor risk acceptance;

XI. **Confidence in experts/regulators**
 A. experts/regulators are (thought to be) substantively competent;
 B. experts/regulators are (thought to be) using optimal strategies;
 C. experts/regulators are (thought to be) serving the subject's (group) interests.

Source: Vlek and Stallen (1980)

INDIVIDUAL AND COMMUNITY DIFFERENCES WHICH AFFECT PERCEPTION OF RISK

Based on a review of the literature, Covello (1983) concludes that studies of risk perception suffer from several shortcomings. One of these weaknesses is that few researchers have analyzed the effects of organizational and social variables on risk perception. He writes that most studies assume that risk perceptions can be explained by the psychological characteristics of the individual and the extent to which a person can correctly interpret technical information. However, it is clear that differences in people's perceptions of risk cannot be explained solely by differences in psychological makeup and misinterpretation of technical data. Other influences are obviously at work.

Recently, researchers have looked at other variables which have been hypothesized to influence perceptions of risk. Research on some of the more important of these variables is briefly summarized below. One should be aware of the shortcomings of many of these studies, however. First, many are descriptive case studies with little empirical data on which to base conclusions. Second, many of the studies which are more empirical in nature have used data collected from a nonrandom subsample of the population. Typically these have included litigation groups or community action groups. Third, since it is difficult to anticipate disasters, most of the studies have been post-hoc with little or no baseline data. This does not mean that what is summarized lacks validity, only that many of the relationships proposed still need further testing.

Cultural Factors

Cultural beliefs are also important in determining perception of risk. These beliefs may weight judgements of potential risk in the absence of convincing information. Lazarus (1966) suggests that general belief systems are particularly important in appraising threats which are ambiguous.

Religious beliefs can cause followers to ignore risks altogether. Those who believe that their fate is in the hands of a "higher power" for example, may resist objective assessments of risk and may reject preventative or remedial actions (Torry, 1979).

Cultural factors are important in determining how a society perceives and adapts to hazards. The characteristics of a culture and the level of technology it possesses often decides how disasters are dealt with. Burton

(1972) writes that there are three archetypical patterns of response to hazards. The first of these, labeled "folk or preindustrial," relies heavily on adjustments at the small group or individual level. In addition, preindustrial responses are often connected with the invocation of magical powers or sanctions. While the preindustrial response is effective in reducing the impacts of minor hazards, the cultures which rely on this archetypical response typically regard the devastating effects of major disasters as matters of fate.

Modern or industrial responses to hazards rely heavily on science and technology to control the environment. Adjustments to disaster often involve large scale responses and high levels of capital. As Burton points out, however, technology is often successful in controlling small environmental variations, but may fail to protect against major cataclysms. Moreover, reliance on technology leads to a false sense of security which precludes taking other "folk" precautions. For example, building flood control dams may tempt people to build homes on floodplains.

Burton proposes that the optimal adaptation strategy is the "Post-industrial or comprehensive" responses to hazards. This strategy is characterized by a rational mix of technological and behavioral adjustments. Without the proper behavioral adjustments, he suggests, total reliance on technology can actually increase damage and loss.

Most Americans believe that hazardous waste facilities are threats. A 1980 public opinion poll found that over 50 percent of those surveyed would actively protest the siting of a hazardous waste facility within one hundred miles of their homes. Moreover, they said that they would protest even if disposal could be done safely and if the site was regularly inspected for potential problems. Similarly, 50 percent of the respondents suggested that they would want to move away from their homes if such a facility was built (Resources for the Future, 1980). A poll of American environmental groups shows that the risks surrounding the disposal of hazardous wastes are their number one concern (Freudenberg, 1984).

Cultural biases can also affect the way risks of the same magnitude are managed (Wilson and Crouch, 1987). For example, most Americans perceive hazardous wastes to be unique threats. Burns (1984) writes that "Hazardous waste is viewed as an entity rather than a subset of solid waste. Waste materials are seen as more hazardous than the chemically similar raw materials. While citizens give scant attention to the hazards of gasoline tank trucks delivering cargo within the city, they see an urgent need to ban transportation of flammable liquid waste materials through their town." (pg. 190.)

Cvetkovich and Earle (forthcoming) introduce the concept of "hazard images." They suggest that hazard images should be thought of as representations of knowledge about a hazard. Such images can be publicly held and are directly influenced by media exposure. These images are often affect-laden and have important influences on the way risks are perceived and responded to. For example, they suggest that "public fears of nuclear power may be partly due to the (erroneous) elicitation of the image of a mushroom-shaped nuclear cloud."

Plausibility of the Threat

Schwartz, White, and Hughes (1985) write that communities respond to a threat's plausibility, hypothesizing that " . . . controlling for all other factors, a highly plausible threat will generate greater concern than an implausible one" (p. 68). Plausibility is influenced by several factors. Scientific evidence which implies a link between a threat and illness, certainly increases plausibility. Even scientific evidence that is designed to calm people can instead alarm them. This is particularly true if experts disagree or are uncertain in their pronouncements. Physical evidence, such as the discovery of a hazardous waste site, also increases the plausibility of an environmental threat. The news media in the community also play an important part in defining the risk and can play an instrumental role in alarming the public (Cvetkovich and Earle; forthcoming). Organizational responses can also influence the plausibility of a threat. If health or government agencies are perceived as unresponsive or concealing facts, community fears are often elevated. "Delayed release of laboratory results, for example, suggests that something is wrong, that they have not been told everything, or that someone is trying to suppress results. The effect is that people begin to believe there really is a serious threat." (Cvetkovich & Earle p. 70.)

Proximity to the Threat

Shippee, Burroughs, and Wakefield (1980) present evidence which suggests that people farther away from the source of danger, yet who can still be affected by it, are more likely to believe in future occurrences of disaster than those who are closest to the potential threat. They refer to

Festinger's (1957) speculations that the primary factor involved in the perceptions of future threat was the availability of physical cues. When physical cues are present, they serve as tangible evidence which may confirm fears of future danger. In contrast, if these cues are not present, it may become necessary to exaggerate actual risks in order to justify the belief in future dangers. This is consistent with Maderthaner, Guttmann, Swanton, and Otway (1978), who found that those who lived half a kilometer from a nuclear reactor rated it as risky as those living ten kilometers away. In contrast, those living 1.4 kilometers away rated the risk as being much higher. Maderthaner et al. suggest that this is because that at 1.4 k. residents are close enough to be aware of the reactor, but are not in frequent visual contact with it.

In their study of the impacts of toxic exposure on residents of Love Canal New York, Fowlkes and Miller (1982) examined the relationship between several demographic factors and the belief that the toxic waste exposure in the community was widespread, dangerous, and had produced unwanted health effects. The study focused on homeowners in three groups: (1) residents from the inner two rings closest to the canal itself; (2) residents relocated from the outer ring; and (3) residents from the outer ring who chose not to be relocated.

By examining geographic variables, the authors found that proximity of one's residence to the actual waste site did not predict one's beliefs. There was, however, an interaction between education and proximity. Less educated residents, closest to the toxic waste landfill, and better educated residents, living farther from the canal, believed that the exposure was widespread, dangerous, and unhealthy

Fowlkes and Miller suggest that this interaction may be due to differences in the way the toxic waste problem was presented to the residents. That is, those closest to the canal received information that defined the situation as very dangerous, while those farther away from the problem were presented information that suggested that the situation was not very widespread.

In a study of the residents of Three Mile Island, Richardson, Sorensen, and Soderstrom (1987) found an inverse relationship between distance from the nuclear facility and variables measuring sensitivity to the threat, concern about its effects, and family disruptions. However, distance from the nuclear facility showed little correlation with fear.

Perry, Lindell, and Greene (1982) found that before it erupted, the perception of threat posed by the Volcano at Mt. St. Helens, Washington, was relatively undifferentiated by distance. Residents twenty-five miles away from the volcano viewed it equally as threatening as those who lived much closer. Perry et. al suggest that this was probably due to the uncertainty about the potential magnitude of the volcano's eruption. As it later turned out, people living twenty-five miles away from the volcano had as much reason for concern as their neighbors who lived closer. When the volcano finally erupted on May 18, 1980, major damage was recorded as far as forty miles away.

Howe (1988) surveyed New York residents who lived within thirteen miles of toxic waste disposal sites. Howe measured both actual and perceived distance from these sites. She found that perceived distance from the facilities was a significant predictor of concern about potential environmental contamination while actual distance was not. The farther from the disposal sites residents perceived themselves to be, the less environmental concern they expressed. While perceived distance was a significant predictor of concern, the amount of variance it explained was less than 4 percent.

Previous Experience with the Threat

Richardson, Sorensen and Soderstrom (1987) conclude that experience with disasters is a major influence on perceptions of risk and loss. The nature of that experience largely determines future sensitivity to the same threat. In the aftermath of an earthquake, for example, victims become extremely sensitive to the possibility of aftershocks. However, repeated exposure to a particular threat might lead to desensitization to the threat which may result in a redefinition of what is acceptable and what is risky. Richardson et al. found that Three Mile Island residents who were frightened and upset by the accident which occurred in 1979, continued to feel threatened by the possibility of future accidents. Sensitivity to the threat posed by TMI was also found to be strongly correlated with individual psychological distress and family disruption.

Age, Length of Residence, and Presence of Children

Age, length of residence, and the presence of young children in the home have been hypothesized to predict perceived risk. Unfortunately, these measures are usually confounded, making it difficult to ascertain which of the variables actually predicts perceived risk. Van Liere and

Dunlap (1980) state that age is usually found to be negatively correlated with environmental concern. Younger people tend to be more concerned about the environment than older people. Fowlkes and Miller (1982) found that older residents, particularly living farther from the Love canal, tend to view the toxic waste problem as being limited. On the other hand, younger residents, particularly those with young children, tend to see the problem as much more serious.

Length of residence, which is also correlated with age, predicted whether people felt that the situation was dangerous. That is, those who had lived in the community longer were less likely to believe that the situation was hazardous. Given a chance to relocate, those who chose to do so had lived in their homes only an average of 8.5 to 10 years, while those who chose to stay had been residents an average of 24 years.

Consistent with these findings, Hamilton (1985) described several demographic factors which predicted increased concern with the contamination of water by toxic chemicals near Milford, New Hampshire. He found that members of more affluent households were more concerned about the pollution problem. Older residents, who had lived in the community longer, tend to see the problem as less severe, whereas women with young children view contamination as particularly serious.

In their study of the residents of Three Mile Island, Richardson, Sorensen, and Soderstrom (1987) used a composite measure called early family life cycle. This measure combined a respondent's age, number of members of the household, and the number of children under age six. They found that families in an early life cycle were more likely to be sensitive to what happened at TMI and to have experienced family disruptions.

Socioeconomic Status

Van Liere and Dunlap (1980) review evidence that social class as measured by education, income, and occupational prestige had a moderate, positive correlation with general environmental concern. They propose that one reason for this is that people in higher social classes have met their basic needs and, therefore, are free to worry about other aspects of life. Thus, environmental concern is a luxury, which can only be enjoyed after basic material needs are met. Other explanations pose the argument that poor people become habituated to polluted, overcrowded conditions, and are unaware of gradual environmental degradation. It should also be noted, however, that while habituation to poor living conditions is a possible explanation, poor people are also more likely to have direct economic ties to major sources of pollution. Specifically, the poor are more likely to have jobs in agriculture, heavy industry, and mining which produce hazardous by-products, therefore environmental concern or activism may actually threaten their jobs.

Perhaps it is because poor people are reluctant to protest (or are ineffective protesters) that their communities are targeted for hazardous waste facilities. This is consistent with Greenberg and Anderson's (1984) study of thirty-one waste sites in the State of New Jersey where it was found that communities with hazardous waste sites had larger proportions of younger people, older people, blacks, and foreign-born nationals than communities without waste sites. In addition, the authors found that the communities with the *worst* waste sites had a greater proportion of children, elderly, blacks, and a smaller proportion of persons aged 35 to 54, families with incomes exceeding $15,000 in 1969, and residents engaged in professional occupations than communities without hazardous waste sites. In fact, the authors note that socioeconomic status was the most consistent significant difference between communities with hazardous waste sites and those without.

Social Network

In their study of Love Canal, Fowlkes and Miller (1982) reported that younger residents perceived their residences as starter homes and had always expected to move elsewhere in the future. In contrast, older residents did not expect to relocate and were part of established neighborhoods and other social networks. These social networks also were predictive of those who relocated and those who did not believe that exposure from Love Canal was dangerous thereby choosing to remain in place. Those who chose not to relocate had fairly private life styles and were rather independent. The fact that many of their neighbors chose to relocate had relatively little effect on their lives.

> These "Non-believers" espouse a highly individualistic and meritocratic set of values. They are defenders of the status quo, and subscribe to the view that life in present day America is inherently and pervasively risky. Accordingly they hold that the major burden of responsibility legitimately resides with each family to secure the information and resources necessary to safeguard its own welfare (Fowlkes and Miller, 1982, p. 96).

Fowlkes and Miller characterized the "believers," those who felt that Love Canal was dangerous, as more sociable residents. They were also likely to hold views that the welfare of the individual and the community are associated. In addition, the believer's expectations were that the government should protect these interests above the interests of "the industrial order."

Further contrasting believers and nonbelievers, the analysts found that believers: (1) had direct experience with the chemicals which threatened the area; (2) sought information by attending meetings; (3) sought answers from officials; and (4) pushed for testing and other ways to clarify an otherwise ambiguous situation. Nonbelievers, in contrast: (1) relied on the experiences of those close to them; (2) were generally removed from direct sources of information; and (3) defended their views that the chemical problem was limited by attempting to discredit those who believed otherwise.

Kinship with Facility Workers

Consistent with Festinger's (1957) cognitive dissonance theory, Fowlkes and Miller (1982) found that those who were employed by the local chemical concerns tended to minimize the toxic waste problems in Love Canal. Similarly, Dew, Bromet, and Schulberg (1987) reported that of several groups interviewed, workers at Three Mile Island had the least negative perceptions concerning safety at nuclear facilities. It has been suggested that through interaction, those related to facility workers would also tend to minimize potential risks.

Perceived Benefits

Hadden and Hazelton (1980) suggest that public objections to hazardous waste facilities stem from an essential disparity in benefit/risk calculations by potentially affected individuals. Because they are asked to endure high personal costs (in the form of increased risk) while the general population shares in the benefits, local residents may oppose the siting of facilities (Portney, 1986). O'Hare (1977) proposes that remuneration should be offered to local residents to compensate for bearing increased costs.

Some studies, such as Hughey, Sundstrom and Lounsbury (1985); Starr (1966); Zautra, Bachrach and Cofresi (1987) have found strong inverse relationships between perceived risk and perceived benefits. Other studies suggest that finding appropriate economic incentives may be difficult. Portney (1986) surveyed attitudes toward the siting of a hazardous waste facility in Massachusetts. Over three hundred randomly selected residents from five communities were interviewed. The results revealed that nearly 62 percent of the respondents opposed the siting of a hazardous waste facility in their neighborhood or state. Moreover, none of the economic incentives proposed to compensate local residents appeared to significantly influence opinions. These incentives included having the developer pay all property taxes for ten years and having the developer repave the city's streets, among others. Portney concludes "... the focus of the theory of compensation on economic incentives probably underestimates the role or magnitude of peoples' risk assessments in the overall benefit/risk calculation. In other words, it would appear that, in general, people assess the risks associated with living near a hazardous waste treatment facility as being so great that virtually no reasonable amount of compensation, by itself, can have much impact (p. 5)."

Prior or Current Health Problems

Often, the presence of health problems in a community first signals that a toxic exposure problem exists (DiPerna, 1985). In addition, once a toxic exposure problem has been uncovered, the occurrences of illnesses and unusual medical conditions quickly become attributed to the contamination (Edelstein, 1988). The discovery of toxic contamination also offers a causal schema for unexplained occurrences, strange smells, tastes, and other experiences. For many, as in the Legler and Woburn cases (Edelstein 1982; DiPerna, 1985), notification of the existence of toxic chemicals in the environment provides a kind of confirmation of suspicions surrounding patterns of illness.

Fowlkes and Miller (1982) reported that health experiences were predictive of whether a resident felt threatened by the toxic situation in Love Canal. Those residents who contracted diseases or illnesses, which were not out of the ordinary, were curable, or controllable, view the hazardous waste problem as less widespread or threatening than those who had suffered from "a strong pattern of unpredictably recurring, debilitating and diagnostically elusive illnesses" (p. 104).

In particular, those residents with young children tend to take health implications associated with toxic exposure seriously. Levine and Stone

(1986) conducted two waves of interviews in the Love Canal area to investigate differences in resident's perceptions regarding the Love Canal situation. The first set of interviews took place at the end of 1978, the second set in the spring of 1979. After the first set of interviews, Levine and Stone found several interesting correlations between social background variables and specific perceptions. With regard to health status, young heads of households, families with young children, and homeowners (particularly those living closest to the canal) were found to be most likely to report poor health in their families over the preceding years.

It is not surprising to find that there is a strong correlation between perceived health status and psychological distress (Tessler & Mechanic, 1978). Stressful environmental conditions have long been recognized as causal contributors to both mental health and physical problems (Cassel, 1970; Jemmott and Locke, 1984). It is not surprising, therefore, to find that disaster victims in Times Beach, Missouri, were significantly more likely to report their health as fair or poor one year after the disaster compared with those who were indirectly exposed or unaffected by disaster (Smith, Robins, Przybeck, Goldring, and Solomon, forthcoming).

SUMMARY AND FUTURE DIRECTIONS

The current state of knowledge of the factors which influence perception of risks is inadequate. Studies on risk perception appear in the literatures of many disciplines. Despite the interests of, and importance to psychologists, anthropologists, sociologists, risk managers, regulators, and policy makers, the study of risk perception has not been characterized by programatic research. In addition, while several theoretical conceptions of risk exist, they have gone largely untested, and are not very useful for prediction. Psychometric methods of research have yielded interesting results, but they have generally been conducted in the laboratory and have dealt with broad classifications of hazards (comparing the risks of nuclear power with skiing, for example). In contrast, field studies have often examined perceptions of risk associated with a specific threat, such as a particular nuclear power plant or hazardous waste disposal site. Unfortunately, these studies generally use small samples of litigants or members of community action groups. Moreover, these studies have tended to focus on a small number of variables (usually demographic), thought to be related to risk. In summary, much of the theory surrounding risk perception has failed to influence empirical studies, and most of the empirical studies have failed to advance existing theories. As a result, neither theory nor empirical research has been sufficient to properly inform professionals who must deal with the public's perceptions of risk. Future research directions should include methodologically sound field studies which examine the relationship of cognitive, emotional, and demographic variables to specific hazards, using a large number of randomly selected subjects likely to be affected by that threat. Without studies such as these, our knowledge of what influences risk perception will remain largely descriptive, with little predictive power and limited utility.

REFERENCES

Ames, Bruce N., Magaw, Renae, and Gold, Lois S.: Ranking possible carcinogenic hazards. *Science, 236:* 271–280, 1987.

Baum, Andrew, Fleming, Raymond, and Davidson, Laura M.: Natural Disaster and Technological Catastrophe. *Environment and Behavior, 15:* 333–354, 1983.

Bolin, Robert: *Psychosocial impacts of natural and technological disasters.* Paper presented at the Sixth Irvine Symposium on Environmental Psychology, University of California, Irvine, May, 1987.

Burns, Marion E.: Striking a reasonable balance. In Harthill, M. (Ed.): *Hazardous Waste Management: In Whose Backyard?* Boulder, Westview Press, for the American Association for the Advancement of Science, 1984.

Burton, Ian: Cultural and personality variables in the perception of natural hazards. In Wohlwill, Joachim F. and Carson, Daniel H. (Eds.): *Environment and the Social Sciences: Perspectives and Applications.* Washington, DC, American Psychological Association, 1972, pp. 184–195.

Cassel, J.: Physical Illness in Response to Stress. In Levine, Sol, and Scotch, Norma A. (Eds.): *Social Stress.* Chicago, Aldine, 1970.

Cohen, Bernard L., and Lee, I-Sing: A catalog of risks. *Health Physics, 36:* 707–722, 1979.

Couch, Stephen, and Kroll-Smith, J. Stephen: The chronic technical disaster: Toward a social scientific perspective. *Social Science Quarterly, 66:* 564–575, 1985.

Covello, Vincent T.: The perception of technological risks: A literature review. *Technological Forecasting and Social Change, 23:* 285–297, 1983.

Cvetkovich, George, and Earle, Timothy C.: Classifying hazardous events. *Journal of Environmental Psychology, 5:* 5–35, 1985.

Cvetkovich, George, and Earle, Timothy C.: Hazard images, evaluations and political action—the case of toxic waste incineration with implications for hazard communication. In Kasperson, Roger E., and Stallen, P. (Eds.): *Communicating health and safety risks to the public: International Perspectives,* forthcoming.

Dew, Mary A., Bromet, Evelyn J., and Schulberg, Herbert C.: Application of a temporal persistence model to community residents' long-term beliefs about the

Three Mile Island nuclear accident. *Journal of Applied Social Psychology, 17:* 1071–1091, 1987.

DiPerna, Paula: *Cluster Mystery: Epidemic and the Children of Woburn, Mass.* St. Louis, C.V. Mosby, 1985.

Douglas, Mary, and Wildavsky, Aaron: *Risk and culture.* Los Angeles: University of California Press, 1982.

Earle, Timothy C., and Lindell, Michael K.: Perception of Industrial Risks: A free-response approach. In Waller, Robert A., and Covello, Vincent (Eds.): *Low Probability/High Consequence Risk Analysis: Issues, Methods and Case Studies.* New York, Plenum, 1984, pp..

Edelstein, Michael R.: *The social and psychological impacts of groundwater contamination in the Legler section of Jackson, New Jersey.* Report to the law firm Kreindler & Kreindler, 1982.

Edelstein, Michael R.: *Contaminated Communities: The social and psychological impacts of residential toxic exposure.* Bolder, Westview Press, 1988.

Elliott, Michael L. Poirier: Improving community acceptance of hazardous waste facilities through alternative systems for mitigating and managing risk. *Hazardous Waste, 1:* 397–410, 1984.

Erikson, Kai T.: *Everything in its path: Destruction of community in the Buffalo Creek flood.* New York, Simon and Schuster, 1976.

Festinger, Leon: *A theory of cognitive dissonance.* Stanford, Stanford University Press, 1957.

Fischhoff, Baruch, Slovic, Paul, Lichtenstein, Sarah, Read, Stephen, and Combs, Barbara: How safe is safe enough? A psychometric study of attitudes towards technological risks and benefits. *Policy Sciences, 9:* 127–152, 1978.

Fischhoff, Baruch, Watson, Stephen R., and Hope, Chris: Defining risk. *Policy Sciences, 17:* 123–139, 1984.

Fowlkes, Martha, and Miller Patricia: *Love Canal: The Social Construction of Disaster.* Report to the Federal Emergency Management Agency, 1982.

Freudenberg, Nicholas: Action for environmental health: Report of a survey of community organizations. *American Journal of Public Health, 74:* 444, 1984.

Gardner, Gerald T., Tiemann, Adrian R., Gould, Leroy C., DeLuca, Donald R., Doob, Leonard W., and Stolwijk, A. J.: Risk and benefit perceptions, acceptability judgments, and self-reported actions toward nuclear power. *The Journal of Social Psychology, 116:* 179–197, 1982.

Greenberg, Michael R., and Anderson, Richard F.: *Hazardous Waste Sites: The Credibility Gap.* New Brunswick, The Center for Urban Policy Research, 1984.

Hadden, Susan G., and Hazelton, Jared: Public policies toward risk. *Policy Studies Journal, 9(1):* 109–117, 1980.

Hallman, William K.: *Risk Assessment, Attribution of Responsibility, and Psychological Distress in a Community Living Near a Hazardous Waste Facility.*

Hamilton, Lawrence C.: Who cares about water pollution? Opinions in a small-town crisis. *Sociological Inquiry, 55:* 170–181, 1985.

Hohenemser, C., Kates, R. W., Slovic, P.: The nature of technological hazard. *Science, 220:* 378–384, 1983.

Howe, Holly L.: A comparison of actual and perceived residential proximity to toxic waste sites. *Archives of Environmental Health. 43:* 415–419, 1988.

Hughey, Joseph B., Sundstrom, Eric, and Lounsbury, John W.: Attitudes toward nuclear power: A longitudinal analysis of expectancy-value models. *Basic and Applied Social Psychology, 6:* 75–91, 1985.

Jemmott, John B. III, and Locke, Steven E.: Psychosocial factors, immunologic mediation, and human susceptibility to infectious diseases: How much do we know? *Psychological Bulletin, 95:* 78–108, 1984.

Johnson, Eric J., and Tversky, Amos: Representations of perceptions of risks. *Journal of Experimental Psychology: General, 113:* 55–70, 1984.

Lazarus, Richard S.: *Psychological stress and the coping process.* New York, McGraw-Hill, 1966.

Lazarus, Richard S., and Folkman, Susan: *Stress, Appraisal and Coping.* New York, Springer, 1984.

Lazarus, Richard S., and Launier, R.: Stress-related transactions between person and environment. In Pervin, L. A. and Lewis, M. (Eds.): *Perspectives in interactional psychology.* New York, Plenum, 1978.

Levine, Adeline G., and Stone, Russell A.: Threats to people and what they value: Residents' perceptions of the hazards of Love Canal. In Lebovits, A. H., Baum, Andrew, and Singer, Jerome E. (Eds.): *Advances in Environmental Psychology. Volume 6 Exposure to Hazardous Substances: Psychological Parameters.* Hillsdale, Lawrence Erlbaum, 1986, pp. 109–130.

Maderthaner, Rainer, Guttmann, Giselher, Swaton, Elisabeth, and Otway, Harry J.: Effect of distance upon risk perception. *Journal of Applied Psychology, 63:* 380–382, 1978.

Mileti, Dennis: Human adjustment to the risk of environmental extremes. *Sociology and Social Research, 64:* 327–347, 1980.

Miller, Patricia Y., and Fowlkes, Martha R.: In defense of "man-made" disaster. *Natural Hazards Observer, 8:* 11, 1984.

O'Hare, Michael: Not on my block you don't—Facilities siting and the importance of compensation. *Public Policy, 25:* 407–458, 1977.

Paterson, Randolf J., and Neufeld, Richard W. J.: Clear Danger: Situational Determinants of the Appraisal of Threat. *Psychological Bulletin, 101:* 404–416, 1987.

Perry, Ronald W., Lindell, Michael K., and Greene, Majorie R.: Threat perception and public response to volcano hazard. *The Journal of Social Psychology, 116:* 199–204, 1982.

Philipson, L. L.: Risk evaluation: a review of the literature. In Covello, Vincent T., Menkes, Joshua, and Mumpower, Jeryl (Eds.): *Risk evaluation and management.* New York, Plenum, 1986, pp. 319–333.

Portney, Kent E.: *The role of economic factors in lay perceptions of risk: A look at citizen attitudes with special reference to hazardous waste treatment facility siting.* Medford, Tufts University, Lincoln Filene Center for Citizenship and Public Affairs, 1986.

Reissland, John, and Harries, Vaughan: A scale for measuring risks. *New Scientist:* 809–811, 1979.

Resources for the Future: *Public Opinion on Environmental Issues: Results of a National Public Opinion Survey.* Washington, DC, 1980.

Richardson, Brad, Sorensen, John, and Soderstrom, E. Jon: Explaining the social and psychological impacts of a nuclear power plant accident. *Journal of Applied Social Psychology, 17:* 16–36, 1987.

Rowe, William D.: *An Anatomy of Risk.* New York, Wiley, 1977.

Schwartz, Steven P., White, Paul E., and Hughes, Robert G.: Environmental Threats, Communities, and Hysteria. *Journal of Public Health, 14:* 58–77, 1985.

Shippee, Glenn, Burroughs, Jeffrey, and Wakefield, Stuart: Dissonance theory revisited: Perception of Environmental hazards in residential areas. *Environment and Behavior, 12:* 33–51, 1980.

Sims, John H., and Baumann, Duane D.: Educational programs and human response to natural hazards. *Environment and Behavior, 15:* 165–189, 1983.

Slovic, Paul, Fischhoff, Baruch, and Lichtenstein, Sarah: Rating the risks. *Environment, 21:* 14–20; 36–39, 1979.

Slovic, Paul, Fischhoff, Baruch, and Lichtenstein, Sarah: Facts and fears: Understanding perceived risk. In Schwing, Richard C., and Albers, Walter A. (Eds.): *Societal Risk Assessment: How safe is safe enough?* New York, Plenum, 1980, pp. 181–216.

Slovic, Paul, Fischhoff, Baruch, and Lichtenstein, Sarah: The psychometric study of risk perception. In Covello, Vincent T., Menkes, J. and Mumpower, J. (Eds.), *Risk Evaluation and Management.* New York, Plenum, 1986, pp. 3–24.

Sjoberg, Lennart: The risks of risk analysis. *ACTA Psychologica, 45:* 301–321, 1980.

Smith, Elizabeth, Robins, Lee N., Przybeck, Thomas R., Goldring, Evelyn, and Solomon, Susan D.: Psychosocial consequences of a disaster. In Shore, James (Ed.): *Disaster stress studies: New methods and findings.* Guilford, 1986.

Starr, Chauncey: Social Benefit versus technological risk. *Science, 165:* 1232–1238, 1969.

Tessler, Richard, and Mechanic, David: Psychological distress and perceived health status. *Journal of Health and Social Behavior, 19:* 254–262, 1978.

Torry, William I.: Anthropological studies in hazardous environments: Past trends and new horizons. *Current Anthropology, 20:* 517–540, 1979.

Tversky, Amos, and Kahneman, D.: The framing of decisions and the psychology of choice. *Science, 45:* 273–300, 1981.

Van Liere, Kent D., and Dunlap, Riley E.: The Social Bases of Environmental Concern: A review of hypotheses, explanations and empirical evidence. *Public Opinion Quarterly, 44:* 181–197, 1980.

Vlek, Charles, and Stallen, Pieter-Jan: Rational and personal aspects of risk. *ACTA Psychologica, 45:* 273–300, 1980.

Wilson, Richard, and Crouch, E. A. C.: Risk assessment and comparisons: An introduction. *Science, 236:* 267–270, 1987.

Zautra, Alex J., Bachrach, Kenneth, and Cofresi, Ana: *Assessment of Two Communities Adjacent to the Planned Hazardous Waste Treatment Facility.* Final report to the Arizona Department of Environmental Quality and Department of Health Services, 1987.

Chapter 4

COMMUNITY RESPONSES TO EXPOSURE TO HAZARDOUS WASTES

KURT FINSTERBUSCH

In this chapter the general findings about community responses to exposure to hazardous wastes will be summarized and related to the literature on social movements. The summary of general findings is largely based on our earlier systematic case review of 25 cases, so the first section of this chapter describes our systematic case review methodology. In the second section the general patterns for the five phases of a toxic waste contamination episode are presented and illustrated in the case of the Velsicol dumpsite in Tennessee. In the third section the behavior of the victims and the community in the 25 cases is described in more depth and two cases, Love Canal and Stringfellow Acid Pits, are analyzed in detail. The relevant findings of Nicholas Freudenberg's survey of 110 community organizations involved in environmental health issues are also summarized. The final section examines the fit of the community responses to toxic waste episodes to the literature on social movements.

METHODOLOGY

Knowledge is the systematic loss of information. It is achieved through cycles of information gathering and distilling. Both the gathering and distilling must be selective and one of the best ways to be selective is to be systematic. When information is gathered systematically in non experimental settings, only the most crucial data are recorded and almost all of the available data is ignored. Only an infinitesimal amount of a respondants' knowledge is tapped, only a tiny fraction of all events are observed, and only a fraction of what is observed is recorded. Nevertheless, billions of bits of the information are gathered which must be distilled to become knowledge. Statistics are computed, patterns are observed, theories or hypotheses are tested and retested and gradually the body of knowledge grows. Through systematic gathering and distilling knowledge is developed that enables us to predict with increasing accuracy how a dependent variable will change when we know how a few independent variables will change. This knowledge sometimes even allows us to improve situations.

The generation of knowledge in a new research area which involves nonexperimental settings usually begins with a case study and expands as other case studies are added that present new configuations of factors. The cases are then analyzed comparatively to identify and explain the commonalities and differences. Where feasible quantitative methods greatly facilitate the analysis of the available cases. The next step for a new research area is the focussed and systematic study of a random sample of cases. This type of study lacks the breadth of the more holistic case studies but provides greater precision and analytical control. In the area of hazardous waste exposure episodes this later step has not been taken. The field is dominated by case studies and qualitative reviews of case studies. To more systematically distill the knowledge from available case studies we have applied a systematic case review methodology to 25 toxic waste contamination episodes affecting humans (Finsterbusch, 1987b). The systematic case review entails three steps:

1. Select a set of comparable cases for review;
2. Develop an information questionnaire and score one questionnaire for each case on the basis of a review of available case materials; and
3. Statistically analyze the results.

These procedures force the analyst to determine early in the review process what information to use in the comparative analysis and to devise the questionnaire accordingly. Then the questionnaire is used to convert the qualitative information of the case materials into dummy variables or scores on scales for comparative statistical analysis. In the process much unique or idiosyncratic information is lost, but the general patterns and causal relationships can be more accurately assessed. The more normal summary procedure is the qualitative review which is more susceptible to judgmental errors (bias) as demonstrated in judgmental psychology (Mumpower and Anderson, 1983).

The first and third steps above should involve standard social science techniques whenever possible. The sample of cases should be selected randomly as the statistical procedures require so that the results can be generalized to the full population of cases with full confidence. The cases for our study, however, were not randomly selected. We used all available cases that were reasonably well documented and accessible in published materials. Published case studies, however, are ones that have achieved some notoriety. Nevertheless, people who have observed other cases have found our summary analysis generally compatible with their observations, and where their case differs special factors seem to explain the deviation.

The second step in the systematic case review method is not standard. In the standard review procedure the analyst examines the case materials and intuitively identifies the general patterns in the cases. Sometimes quantitative and nominal variables are recorded for each case and frequencies or means are reported, but most variables are qualitatively assessed because they are presented qualitatively in the case materials. The systematic case review method, however, translates qualitative case materials into ordinal scales for quantitative analysis through subjective scorings by coders. When two or more coders are used, measures of intercoder reliability can be computed. In this study only one coder, the author, scored the cases, so the reliability of the scorings cannot be assessed. This shortcoming will be rectified in future work.

The systematic case review method can produce very useful results, but it is far from a perfect method. Case materials often say little about many of the variables in the information questionnaire, and often it is difficult to score different types of cases comparably on many of the dimensions. The case materials have their own deficiencies which add error to the systematic case review. The results must be viewed with some caution, therefore, and hopefully the more important findings will be retested with even a more reliable methodology.

PATTERNS IN TOXIC WASTE CONTAMINATION EPISODES

For analytical purposes it is useful to divide toxic waste contamination episodes into five phases: predisposal, disposal, discovery, reaction, and resolution/outcome. The following common pattern was found in our review of 25 cases (Finsterbusch 1987a, 1987b):

1. The predisposal phase was typified by inadequate and minimal regulatory systems governing the disposal and storage of toxic

wastes. Where regulatory systems existed they were not strongly enforced.

2. In the disposal phase polluters did not use sound disposal practices even though many of them acted legally and used state-of-the-art disposal methods. Even technically and legally acceptable techniques caused off-property contamination and threatened lives and property. Polluters usually knew that their disposal techniques would contaminate surrounding properties or groundwater, but usually were unaware of the extent of the health risks to which they exposed people.
3. In the discovery phase the victims were usually confused and ignorant of the danger they were in. They often detected contamination because of bad odors in the air or tastes of well water, but they did not understand that the contamination could cause sicknesses and eventually death. They complained to authorities and usually received very inadequate responses. The authorities discovered or recognized the contamination before the victims in nine cases, but in the majority of these cases the authorities were slow to recognize the problem and slow to act. They even at times kept information about the contamination from the public. Only in two cases did the media play a *major* role in discovering and broadcasting the problem.
4. The reaction phase found victims complaining and demanding actions from the authorities who usually acted slowly, stingily, and inadequately. The victims, therefore, had to organize and esculate their pressure on the authorities to investigate, stop further pollution, decontaminate and compensate. Their pressure tactics were only partially successful. Meanwhile polluters continued polluting on average 4.1 years after public discovery of the problem.
5. The resolution usually involved the removal of danger (though sometimes with considerable delays) but little removal of the toxins. Waste lagoons were covered, wells closed, toxic emissions lowered to "safe" levels, or people moved off contaminated land. Victims usually received little or no compensation. Polluters were never prosecuted, paid few fines, and in the majority of cases paid no compensation to victims.

The Velsicol Dumpsite Case

The case of the Velsicol dumpsite in Hardeman County, Tennessee (Brown, 1980; Epstein, Brown, and Pope, 1982; Freudenberg, 1984b) will be used to illustrate the above patterns. When Velsicol bought a farm and converted it into a dumpsite in 1964, this action violated no law. According to Epstein, Brown and Pope (1982):

> The farm's conversion into a dump required no permits, licenses, or public announcements. Tennessee law, at that time, allowed a landowner to do anything with his own land, as long as he didn't obviously bother anyone else. Velsicol did not have to inform any state or local agency that it had established the dump or what it was dumping. No laws specifically prohibited the dumping of any materials on private land, nor had state and local officials any authority to check on the operation. (p. 51.)

Velsicol used a simple and inexpensive disposal technique. A bulldozer would dig a trench about 12 feet deep and fifteen feet wide and trucks would dump their cargo of toxic wastes contained in 55 gallon steel drums or fiber cartons into the trench. The bulldozer would crush the drums and then spread a layer of dirt in the trench. More drums would be dumped, crushed, and covered until the trench was full and covered with a layer of dirt. Since the trench was not lined, the wastes leached into the soil. About 300,000 drums of toxic wastes were disposed in this manner.

The discovery phase was long and complicated. It took six years for responsible authorities to determine that there was a problem after a Memphis County engineer officially questioned in March 1965 the safety of the dump, because it was over the artesian aquifer that served as a water source for the city of Memphis. The Tennessee Stream Pollution Control Branch spot tested and found contamination in several shallow wells on the perimiter of the dump. The state did not inform the residents but did inform the county health officials and called for a U.S. Geological Survey study of the site. The survey was conducted in 1966 and 1967 and found that contamination had almost reached 90 feet down to the local water-table aquifier. The survey concluded, however, that nearby residents were in no danger, because the prevailing movement of groundwaters was in the opposite direction.

Encouraged by the report Velsicol increased its dumpsite area from 20 to 45 acres, and the state had no authority to stop it even though state officials were concerned. In 1970 a USGS hydrologist raised new concerns about the safety of the site.

> These questions resulted in a high-level, closed-door meeting on March 4, 1971. Representatives of the Tennessee Department of Public Health, the USGS, Congressman Ray Blanton, and Velsicol were in attendance. Neither the press nor general public was informed. The USGS frankly admitted that its 1967 study had failed to recognize the extent of the danger from the Velsicol operation, a danger subsequently exacerbated by Velsicol's expanded dumping. Velsicol officials were unmoved and made it clear that they intended to continue dumping. (Epstein, Brown, and Pope, 1982, pp. 54.)

Even though responsible officials knew about the problem in 1971, it was seven more years before the public learned of the danger. Neighboring residents knew that Velsicol was dumping odious wastes on its property from the beginning, because they were bothered by noxious fumes from the open trenches. It was not until 1977, however, that they noticed strong chemical tastes in their drinking water and complained of headaches, nausea, and loss of energy. They asked George Wallace of the County Health Department to examine their drinking water. He tested only for bacteria even though the 1965 report of chemical contamination had been sent to him. Of course, he found no contamination. The neighbors' health continued to worsen, and several were hospitalized for mysterious illnesses. Several more times Wallace tested wells in response to complaints but always tested only for bacteria. The residents also asked the state health department to test their water, but their requests went unheeded for six months. They brought a sample to the State Water Quality Division Laboratories, and the lab found contamination. Further tests were made and finally on March 23, 1978, the state health department and EPA notified them that their water was dangerously contaminated.

The above account reveals a complex tangle of inaction, irresponsibility, unseemly delays, deceit, incompetence, duplicity, and suppression of information, on the one hand, and initiative, persistence, and professional competence, on the other hand. The mixture of these components varies considerably from case to case, but often a complex mixture such as this is to be observed.

The reaction phase for many cases has the victims requesting or demanding that the authorities take appropriate actions. In a minority of cases, including this one, the authorities act before the victims learn of the problem and pressure them. After the high level meeting in March, 1971,

> The Tennessee Department of Public Health tried without success to persuade Velsicol to stop the dumping. Finally, in February, 1972, the state filed a complaint against Velsicol. Velsicol was ordered to halt all dumping, except that of nontoxic wastes . . .
>
> While Velsicol began burning some of its wastes out of state, it nevertheless continued dumping at the Hardeman site at a rate of about four truckloads a day. In spite of persistent complaints from local residents, no enforcement action was taken.
>
> By 1975, Velsicol had finally stopped dumping. The Hardeman site was officially closed and given a superficial cleanup (Epstein, Brown, and Poper, 1982, p. 54).

After the tests of March 1978, the neighbors were told to stop using their water, but alternative water was not provided. Some residents moved away and some hauled drinking water from 20 miles away. Federal funds would pay $240,000 for piping town water to the area, if the town of Toone would underwrite a $30,000 loan, but the town board refused. Some town citizens and the media criticized the decision, and heavy pressure developed for the board to reverse its decision. This it did, but it stipulated that 30 dumpsite area families had to pay $165 each in advance for the water hookup. With great difficulty the money was raised, but six of the checks bounced so the town cancelled the deal.

A congressional committee began to investigate the case and came across a six-week-old EPA report showing carbon tetrachloride in the drinking water of one of the homes at the shocking level of 4,800 ppb. The report had been sent to the Tennessee Department of Public Health which had ignored it. Embarrassed, the department tested further and ordered the contaminated wells capped. The investigation also uncovered several other cases of bungling and agency ineptitude. The major effect of the investigation, however, was to stimulate Velsicol to start to deal with the problem. Velsicol finally stopped denying liability. "Preliminary analysis just received may suggest the former disposal site is the source of contamination . . . " Velsicol lamely announced. Then it paid the $30,000 water hookup charge that Toone was not paying and replaced the contaminated plumbing and appliances of contaminated households. It bought many homes at a high price and then rented them back to the family at low rates. The area residents had filed a $2.5 billion suit against Velsicol back in May, but as Velsicol stepped in and helped them out many signed releases.

The resolution of the problem begins with the capping of the contaminated wells and the provision of piped water to the area. This removes residents from further exposure to dangerous levels of the toxins. The long-term health effects of their past exposure, however, cannot be determined. As of late 1983, the $2.5 billion suit had not been decided. Since both sides have vowed to appeal an adverse decision settlement will take a long time.

The final resolution of the dumpsite itself is unclear. Velsicol capped the site and monitored it until the end of 1983. Then Velsicol, the State, and EPA were to determine what further actions need to be taken, but our sources do not report in detail beyond this point. In 1989, EPA's Information Office reports that the State of Tennessee has taken responsibility for "remediating" the site and that its cleanup is still in progress. EPA estimated in 1979 that complete clean up would cost about $165 million, but that would not clean up contaminated groundwater. Contaminants are moving underground at a rate of 80 to 300 feet per year and moving in the direction of the Toone-Teague wells. Furthermore... "there is now the real possibility that cracks or fissures in the subsurface below the dumpsite may allow some 16.5 million gallons of highly toxic wastes to filter down into the deep artesian aquifer that supplies Memphis with its drinking water." (Epstein, Brown, and Pope, 1982, p. 65.)

COMMUNITY RESPONSE TO CONTAMINATION EPISODES

The responses of the public to exposure to toxic pollution seems to be guided by a tradition of letting the government handle threats to public health. Community responses include much confusion, but some common patterns can be observed. In this section we detail more fully the common patterns of citizen responses which were briefly mentioned in the above description of the discovery and reaction phases, review the survey results of Nicholas Freudenberg, and summarize the public responses to the Love Canal and Strangefellow Quarry episodes.

The 25 Case Study

In 16 of the 25 cases in our study mentioned earlier the victims figure out that something is wrong before the authorities do. The victims usually do not understand what is poisoning them or the danger they are in, but they know that the air or water smells or tastes badly and is stinging their eyes or making them feel dizzy or nauseous. Velsicol's neighbors smelled the fumes from the dumpsite and eventually noticed chemical tastes in their well water. They complained to government officials, asked that the situation be investigated, got false reassurances or no action, continued complaining and eventually were told that their water was highly toxic and they must not drink it any more. This pattern, in every detail, was repeated in Love Canal, Stringfellow Quarry, and eight other cases. In five remaining cases where the victims discovered the problem the response scenarios were much more idiosyncratic. In two cases the citizens did not complain to officials as far as we know. In one case the tests by the authorities never did find dangerous levels of chemicals and in only two cases did the complaining citizens get real help from an official right away.

In eight of the nine cases where the officials discovered the pollution first the citizens were suddenly informed of the problem and the citizens were more passive, at least at first. In six cases the citizens left the government alone to deal with the problem, but in three cases they became dissatisfied with the government actions and demanded more action. The most common pattern, therefore, was for the victims to experience a nuisance, ask the government to investigate and solve the problem, receive inadequate and often incompetent government responses, and pressure for a more adequate government action. The second most common pattern was for the government to find out about the problem before the victims and to take actions like closing wells, stopping further disposal or buying out homeowners while the victims remained relatively passive. A major finding of this study is that *citizen activism is generally a response to government inactivity or failure.*

Another pattern that is quite noticeable in the 25 cases is the important role played by one or more citizens who became persistent, stubborn, and demanding in order to stop the pollution and obtain redress. One neighbor of Velsicol, Nell Grantham, a licensed practical nurse, kept demanding water tests and urged the class action suit against Velsicol. Later she became a leader in a new statewide coalition, Tennesseans Against Chemical Hazards. The spirited leader of the Love Canal residents was Lois Gibbs who now is the director of the nationwide Citizen's Clearinghouse for Hazardous Wastes which helps citizens around the country to organize against toxic waste sites and proposed facilities. The early leader of the fight against the Stringfellow Quarry site was Ruth Kirkby, a nearby resident who was an expert in paleobotany and other earth sciences. She wrote many letters to authorities, made many phone calls and organized the Parents of Jurupa which fought the quarry. Many of the other victim groups had similar leaders and sometimes in low density areas a feisty victim would battle the polluter, and sometimes the authorities, alone.

Study of 110 Community Organizations

Frendenberg's study (1984a) adds to the above picture of community responses to toxic waste contamination episodes. He surveyed 110 community organizations involved in environmental health issues, the most common of which were toxic dumpsites, air pollution, and pesticide spraying. Using a mailed questionnaire he asked about the groups' goals, methods, problems, and accomplishments. Most of the groups were local (55 percent), 30 percent covered a catchment area and 9 percent were state or regional. Their major concern was health problems, especially cancer, and 48 percent believed that they had documented, at least to their own satisfaction, that the exposure had caused illnesses. Many also addressed other issues such as conservation, taxes, and peace, and 42 percent were affiliated with some larger environmental, citizen action, or other group. "Nearly half (46 percent) the groups were formed because concerned citizens became alarmed or angry about suspected health hazards" (p. 445) and thus are like the groups studied in this chapter.

Freudenberg summarizes his findings on the groups' goals activities and accomplishments as follows:

> The goal identified most frequently was the correction of the specific hazard facing their community. To achieve this aim, groups engaged in a variety of activities including research, public education, demonstrations, lobbying, and legal action. Most groups reported extensive and helpful interactions with scientists or health professionals. Respondents reported problems in obtaining information from local health officials, other government agencies, and industry. A majority of the groups rated their efforts at environmental health protection as being very or somewhat successful. (p. 444.)

An interesting finding is the amount of effort that often went into obtaining information and the extent government agencies blocked their efforts.

> A group in the State of Washington filed 200 separate Freedom of Information Act requests to obtain data on the herbicides being sprayed

in that area. A Kentucky group turned to the staff of its congressional representative for information on nerve gas stored on a nearby military base, while a group in New Jersey used records of the U.S. Securities and Exchange Commission to investigate previous environmental violations of the company planning to build a toxic chemical waste facility in its city.

Nearly nine out of every ten groups (88 per cent) perceived obstacles to obtaining information. Almost half (45 percent) claimed that government agencies blocked their learning process. (p. 445.)

The Love Canal Case

The next data source for studying community responses to contamination are the reports on the famous Love Canal dumpsite. In 1942, Hooker Chemical began dumping chemical wastes in an abandoned unfinished canal in a sparcely populated area of Niagara Falls. By 1953, it had filled the canal with highly toxic chemicals and capped it off with clay and covered it with grass. By then the area had built up considerably, and the City of Niagara Falls condemned the property adjacent to the canal to build an elementary school. Hooker agreed to sell Love Canal to the school board for $1 provided the deed absolve Hooker of all future liabilities for the contents of the dump. The school was built, and the kids used the canal area as a playground. The area continued to develop, but problems with the toxins also began to develop. Children suffered chemical burns from playing in the area, odors increased, and leaking chemicals ruined yards and basements. People began to have a wide variety of health problems, especially miscarriages, deformed babies, and ill children. Over time the canal filled with water and the chemicals overflowed and leached, especially along the path of filled in swales. Residents' complaints finally forced the city to do a study which reported in 1977 that chemicals were contaminating the surrounding neighborhood. The residents organized, learned of each others' health problems, petitioned the close of the school, and finally demanded that they be relocated and compensated. The state studied the situation and advised the county to close the school and declare an emergency. The county refused. The state then ordered the school closed, evacuated some residents, and purchased their homes. Remaining endangered residents had to fight long and hard against city and state opposition for compensated relocation. Eventually the federal government moved 550 more families. The toxins have not been removed, but in 1982 the EPA Superfund spent $7 million to prevent the spread of the toxins. (Epstein, Brown and Pope, 1980, Frendenberg, 1984b; Levine, 1982; Gibbs, 1982.)

Love Canal is the most thoroughly examined case of toxic waste contamination in a community. It is both typical and unusual. It is typical because it fits into the most common pattern of community-government responses and interactions. It is unique in its *extreme* values on each type of response and interaction as discussed above. First, the Love Canal residents discovered the problem and became convinced of the danger way before the government agencies did. Second, they complained over a long period of time. Third, the government officials were extremely slow to respond, refused to admit there was a dangerous problem, were very misleading, and were extremely reluctant to do much about it. Fourth, the Love Canal residents became extremely well organized, active, knowledgable. They were very astute political strategists and effective in using the media. Fifth, they were extremely angry, dissatisfied, aggressive, confrontational, and demanding. Sixth, they were led by an effective leader, Lois Marie Gibbs. Finally, they eventually succeeded in obtaining an exceptionally favorable government response, i.e., buyout and relocation of all area residents who wanted to move. The one way residents deviated from the modal pattern was their ability to find out about the problem and become much less dependent on government testing and information. Nevertheless, this example serves as an ideal model for the general pattern of community responses and interactions with government. Levine's (1982, p. 11) pithy summary of the early stage of this episode could be applied to many other cases. "And so it had started: the company dumping wastes; the residents suffering from their proximity to the chemicals—though they did not yet fully realize it; the regulations obsolete; and the regulators uninterested."

In addition to the general lessons of Love Canal, this case provides many special lessons on community responses to contamination episodes. First, residents complained about the odors from the canal and the burns and injuries their children received from playing with materials from it, but they did little more than complain to authorities and most residents did not even do this. Until 1977 most residents displayed two characteristics: *passivity* and *dependency*. They remained essentially inactive and depended on authorities to deal with the problem.

The second special lesson from Love Canal *is the importance of the leader*. Passivity and dependency characterized most of the impacted residents until 1978 when Lois Gibbs went around the neighborhood with her petition to close the elementary school on the edge of the canal. Before this Karen Schroeder helped organize the first group dealing with Love Canal. It was a small group that sought tax abatement and hired a lawyer. The organization of the neighborhood, however, was largely the outcome of the work of Gibbs. She knocked on doors for four or five weeks talking to people and finding out their problems with the canal. She was getting worn out so she asked Debbie Cerrillo to help her with the canvassing and soon they organized the Love Canal Homeowners Association. When two announcements were made on August 2, 1978 that the elementary school would be closed (Gibbs' original objective) and that the State Department of Health recommended that pregnant women and children under age two should move out of the neighborhood, the residents became very troubled and a crowd formed at the school. Gibbs spoke to the crowd and was more or less accepted as its leader. The next day 600 residents attended a meeting of state, county, and local officials. The officials offered a modest amount of help and were shocked at the crowd's response. Lead by Gibbs the residents demanded much more. "We want out! We want out!" they chanted. They were no longer passive and dependent. They had started to fight. Gibbs and others learned to skillfully use the media and to wrest control of events from the officials, and they accomplished extraordinary results against tremendous odds. What was the secret of the residents' success? The answer is organization and leadership.

Third, *the media* has a constitutional mandate to make government officials accountable and to inform the public about things that affect them. When the media carries out its mandate, government officials respond more and dictate less. In Love Canal the media made government officials more responsive to the pleas of the residents. Governer Carey was running for reelection, and national media attention forced him to use emergency funds to buy 239 houses closest to the Canal and relocate the residents. Later, national media attention forced President Carter to buy out and relocate 710 more families.

Fourth, *science can become political.* The subtitle of Levine's book is *Science, Politics, and People* because science became mixed with politics and politics became mixed with science. Levine documents the role of the various health and toxicity studies in the Love Canal episode and the development of the residents' mistrust of the Department of Health researchers. They, therefore, conducted their own study that found adverse health effects of the canal. Experts criticized the methodology of their study, but they in turn found good reasons to criticize the large Department of Health health assessment study. Politically tinged scientific debates are common in environmental conflicts and Love Canal exemplifies this rule.

Finally, the interactions between officials and anxious residents are often unsatisfactory because of the inherent conflict of interest between the two groups and because officials use an inappropriate style. They approach matters cognitively and fail to appreciate the emotional responses of the residents.

> People's reactions to decisions were simply not viewed as appropriate parts of the complex of problems to be addressed. "We deal with physical facts, not with social and political matters," one health official mentioned later. Not only did that agency repeatedly ignore the social facts of people's reactions, however; so did virtually every governmental body that became involved over the next two years. (Levine, 1982, p. 35.)

The Stringfellow Quarry Case

The Stringfellow Quarry is another major hazardous dumpsite episode that has been extensively reported. In 1955, Stringfellow Quarry in Riverside, California became a toxic waste dumpsite for Los Angeles, Orange County, and Riverside itself. The campaign manager for the Riverside County Supervisor operated the site. The toxic wastes were dumped in open pits rising steplike up the granite and sand walls of Pyrite Canyon. Irritating and nauseous fumes frequently bothered nearby neighborhoods and the citizens began complaining. The citizen mobilization increased over the years, but state and local agencies did nothing. Finally, in 1972, Stringfellow announced he was closing the dumpsite, but governments and corporations vainly urged him to keep it open because there were few other places for the toxic wastes to go.

After the site closed it began to deteriorate and the state had to spend $207,600 from 1975 to 1978 to maintain it. Stringfellow claimed that his corporation had no money to pay costs. With time the problem became more acute because water ran down the mountain side into the pits and higher pits overflowed into lower ones. Even the dam at the bottom was threatened. In 1978, 600,000 gallons were released to take pressure off the dam and 275,000 more gallons were released a few days later. Pollution spread downstream, overflowing across farmlands and into residential yards. The problem of the remaining toxins still has not been resolved and the site is on the Superfund list. This was the top priority site in California on the first Superfund list. (Brown, 1980; EPA, 1980; EPA, 1984.)

The Stringfellow episode conforms in most points to the general pattern in the study of 25 cases reported above. The victims knew something was wrong before the authorities did, but they did not understand the extent of the danger. Residents complained to government officials, asked for investigations and corrective actions, were given false assurances and no action, continued complaining, used the media, and eventually helped precipitate significant governmental action on the problem. These results, however, required organization, leadership, intense conflict with authorities, national media attention, and ultimately federal action.

We shall briefly discuss four lessons from the Stringfellow episode which have not been discussed in connection with the studies reviewed above. First, Stringfellow demonstrates (as does Love Canal) that many toxic waste contamination episodes are political footballs. Such episodes can help make or break politicians. It is alleged that EPA's administrator Anne Burford held up money for the site because allotting the money would aid former Governor Edmond Brown in his campaign for the U.S. Senate. Congressman George Brown sponsored a public meeting at the area school that gained him public approval. State Senator Robert Presley held a public meeting on the site, and he has been active in obtaining state funds for containment and partial cleanup on the site. In 1977, Presley and Assemblyman Walter Ingalls introduced and passed a bill in the state legislature which appropriated $370,000 for the cleanup.

An outside group called the Campaign for Economic Democracy (CED), eaded by Tom Hayden, became active in the controversy over the site in 1979. According to CED, Hayden was asked by Governor Brown to investigate the Stringfellow site as a member of the Southwest Border Regional Commission. CED organized dumpsite neighbors into the Concerned Neighbors in Action (CNA). A CED organizor and the original members went door to door throughout the community to sign up members, enlist support, and pressure for a complete cleanup of the site. CNA members showed up in large numbers at the critical hearing of the Santa Ana Regional Water Quality Control Board (RWQCB), which was to determine what to do about the site. CNA's pressure forced the board to decide for the strongest set of actions although it had originally supported a much weaker one. Two weeks later CNA held a public meeting at which Tom Hayden and Lois Gibbs spoke and Congressman George Brown's aide attended.

In sum, politicians in the area paid close attention to the episode. Some of the residents said to interviewers that they thought that some of the politicians, especially Hayden, were being opportunists. The lesson is that this episode provided political opportunities.

The second lesson is *that citizen groups near Stringfellow were supported by national or state environmental action groups.* A national environmental citizens network had evolved by 1980 to the point that considerable outside help was becoming available that could provide knowledge about toxins, dumpsites, health effects, and legal weapons as well as advice on strategies and tactics. Today these outside supports are much more extensive. Local groups are able to obtain legal, technical, political, and personal advice from experienced personnel in several national organizations. One of the newest is the Citizen's Clearinghouse for Hazardous Wastes which Lois Gibbs founded in 1981. It claims that "By January, 1987, we were serving over 1700 grassroots toxics organizations, 400+ multi-issue neighborhood groups, 300+ labor and religious leaders and over 1300 local chapters of established environmental groups" (Citizen's Clearinghouse for Hazardous Wastes, 1987, p. 6).

In the Stringfellow case both state and national environmental groups helped the local effort. Neighborhood residents were organized by an outside organization, Campaign for Economic Democracy, into the Concerned Neighbors in Action (CNA), which demanded a complete cleanup of the site and had considerable influence on the actions that government agencies did take. Lois Gibbs' visit and advice also contributed noticably to the mobilization of the citizens.

The third lesson is the *competence of the citizens* when they do become involved. Often officials and experts underrate the understanding that these citizens have and are shocked when some of their professional reports are torn to schreds in public by "little housewives" and other laymen. This lesson is also well illustrated in the Love Canal episode.

Finally, the Stringfellow case displays *the process of institutionalization* that is occurring throughout the toxic waste field. At the federal level considerable legislation has occurred since 1969 including Superfund which is providing some of the funds for the cleanup of Stringfellow. EPA, however, is trying to spend as little government money as possible, so in 1983 it had the Department of Justice join with California in filing a suit against 31 parties responsible for wastes at the site. Meanwhile EPA has taken emergency measures along with the state to contain the toxins at the site. The final disposition of the site is still being worked out. In the February/March 1989 California Department of Health Services (DHS) newsletter, "Stringfellow Update," DHS and EPA propose a clean-up plan which is estimated to cost about $170 million. It involves "on-site dewatering with extraction and treatment of on-site leachate, in-sites soil-gas extraction, treatability studies, and down gradient plume cleanup" (California Department of Health Services, 1989:6). Alternative plans that are being considered range in costs from $130 million to $740 million. The most expensive plan includes "full excavation of all diggable soil [and] soil treatment followed by redisposal on-site into the excavation" (p. 5). DHS and EPA claim that the preferred plan has community acceptance, because representatives of Glen Avon approved it on September 20, 1988, so the clean-up plan may be selected soon and the clean-up begun.

We note in the Stringfellow case that legislation, administrative procedures, and legal precedents are evolving at the federal level for dealing with toxic waste contamination episodes. EPA has been maligned for footdragging and there is much evidence to support the accusation, but EPA has only slowed the process of institutionalization. It has not stopped the process. Similar institutionalization is taking place at the state (DHS), regional (Santa Ana Regional Water Quality Board), and local levels. Institutionalization is even taking place to some degree at the level of citizen action groups.

COMMUNITY RESPONSES AS A SOCIAL MOVEMENT

Community responses to toxic waste contamination episodes evolves from apathy and reliance on government to mobilization for redress of grievances. They can be classified, therefore, as local social movements. In this section we briefly describe the patterns found in the social movement literature and compare them to the patterns found in the 25 cases. This is a relevant literature for understanding these episodes. They are also a form of community conflict, but the community conflict literature is only marginally helpful (see Coleman, 1957, for the most useful study of community conflict). The political economy literature is also noteworthy (especially Schnaiberg, 1980; Coleman, 1982; Stone, 1980), for these frameworks explain the way authorities act and why.

The social movement literature is currently dominated by the resource mobilization approach which focusses on the constraints on and resources available to the social movement group. It criticizes the social psychology of collective behavior approach as overemphasizing spontaneity and nonrationality of social movement behavior, and it emphasizes the influence of political, economic, and social factors on the rational behavior of participants. It begins by examining the resources of the movement group, the extent of their mobilization and the extent of the mobilized resources of the movement's opponents. The more resourceful group usually wins. The resource mobilization approach recognizes, however, that often people with a mutual interest will not act collectively to serve their mutual interest. In fact, self interested people may act rationally by *not* joining in collective actions to procure public goods unless they are forced to join or are provided special incentives to join. The reason is that they can enjoy the benefits of the public good without incurring the costs of joining the action (Olsen, 1965). For this reason, social movements, cannot depend entirely on rational pursuit of self interests. Rather, people often participate in movements more for some vision of justice or out of obligation to friends than out of self-interest. Not only their interests, but their perceptions and solidarities motivate such participatory behavior. These admissions allow for a rapproachment between resource mobilization and collective behavior theorists.

Other factors affecting movements' effectiveness are the ways that movements are organized, third party contributions, media portrayal, public response to legitimacy claims, and the strategy and tactics of authorities. When these factors are considered together the analysis becomes complex and cumbersome. We, therefore, use as a guide Freeman's (1979) resource mobilization conceptual framework which summarizes the factors that effect the degree and/or success of social mobilization. Freeman's list includes ten types of resources, five types of constraints, nine dimensions of social movement organizations and an unlimited number of environmental factors (see Figure 4-1). The main weakness of Freeman's framework is its relatively static nature so we use Gamson, Fireman and Rytina's (1982) framework as a guide for analyzing the dynamic aspects of social movements. These frameworks will be applied in a basic way to the general pattern of community responses to toxic waste contamination episodes.

Freeman identifies ten resources for mobilizing social movements. Three are tangible resources: money, space, and publicity. For a toxic waste victims' mobilization (TWVM) the money needs are minimal unless the group decides to hire experts or lawyers, space needs are minimal and easily provided, and publicity needs are often large. Publicity includes communication among members and publicity to the public. The former

Dimensions of Social Movements	*Descriptions for Typical Toxic Waste Victims' Mobilization*
I. Resources	Few: time, commitment and publicity are key
1. Money	Little
2. Space	Little: people's homes
3. Publicity	Moderate: inform group, use media for public
4. Expertise	Some: self learning, donated helped, increasing national help
5. Access to networks	Some: environmental networks
6. Access to decision makers	Little: unless politicians use for political gain.
7. Status in polity	Low
8. Status in group	Low at first but grows when authorities fail
9. Time	Often considerable and key
10. Commitment	Often considerable and key
II. Constraints	Relatively few constraints
1. Values	Ideology not important
2. Past experiences	Usually lacking
3. Reference groups	Learn tactics from other demonstrating groups
4. Expectation	Disillusion with authorities lead to acting for themselves
5. Relations with target group	Close: target group are fellow neighbors
IV. External Environment	Low opportunities but also low opposition, appeal to sympathies
1. Structure of opportunities	Formal channels fail and limited alternatives
a. Formal opportunities	Usually used with unsatisfactory results
b. Leverage points	No common pattern. Some use election time to create pressure
c. Creating new opportunities	Help create political reputations
2. Social control encountered	Not much of a problem unless pollutor is a major employer
a. Law enforcement and abuse	Not much of a problem
b. Public opposition	A serious problem when pollutor is a major employer

is essential to mobilization, and the latter may be essential to success. Communication among members of a TWVM is largely by word of mouth and/or by the association newsletter. These are usually readily available and not a significant constraint on mobilization. Both require volunteered time and, in the latter instance, small amounts of money. Publicity to the public requires the use of the local media which can be problematic when the media is strongly influenced by the polluter.

Dimensions of Social Movements	*Descriptions for Typical Toxic Waste Victims' Mobilization*
IV. c. Illegal repression	Not a serious problem
d. Infiltration and disruption	Not a serious problem
e. Ignoring	Often a serious problem
3. Role of bystander public	Sometimes essential to success
a. Solicited as political supporters	Media campaign often essential to success
b. Solicited as members	Not likely because they not share danger
c. Offended so opposing	TWVM's are seldom offensive
d. Undermine their support	Play up victim image in media to win support

Note: Freeman did not specify a list of dimensions in Section III and IV. We have specified the dimensions largely on the basis of her discussion.

Figure 4-1. How the typical toxic waste movement scores on Freeman's Dimensions of Social Movements.

Freeman categorizes three resources as specialized: expertise, access to networks, and access to decision makers. Most TWVMs lack these resources in the beginning but develop them during the course of the episode. Furthermore, nationwide and statewide environmentalist networks have been developing in the last decade which make expertise and networks available to local TWVMs to increase their effectiveness. The TWVMs differ considerably in their access to decision makers but most of them had very little. This often turns out to be the weakest point of the TWVM and the reason why they must create strong pressures on decision makers through the media and political activism.

The remaining four resources are: status in the polity, status among the constituencies of the TWVM, time, and commitment. The two status variables are not as germaine to local TWVMs as they are to national social movements. Nevertheless, we observe that TWVMs have low status in the polity and have to earn status among the victims in order to grow. Time and commitment (the willingness to bear costs) are very important to local TWVMs and are essential to their success. In sum, the key resources for TWVMs are publicity, expertise, access to decision makers, and commitment.

Freeman's five types of constraints are values, past experiences, reference groups, expectations, and relations with target groups. One of the main differences between toxic waste victims' mobilizations (TWVMs) and typical social movements is the low level of definition of these variables for TWVMs and, therefore, their lesser importance. Values, in particular, are relatively unimportant for TWVMs because they mobilize people who are very heterogenious in terms of their values, although members are united in their concern about safety. In terms of tactics and expectations, the many demonstrating groups, from civil rights groups to the antiabortionists, provide reference groups for TWVMs.

Freeman discusses nine dimensions of the social movement organization: associational versus communal, inclusive versus exclusive, isolated versus interacting with the social system, centralization, hierarchy, degree of division of labor, degree of segmentation, number and role of movement organizations, and size. Most TWVMs are communal, inclusive, interacting with social systems, decentralized, nonhierarchical, segmented from other TWVMs, small, have little division of labor, and involve only one or at most two organizations. These organizational features mean that considerable resources must be directed toward maintaining the group and a relatively small amount is directed directly at attaining the groups goals.

Environmental factors are classified into three categories: "(1) the structure of available opportunities for action, (2) social control measures that might be taken, and (3) the effect on bystander publics." (Freeman, 1979, p. 185.) Many factors can be included in each category so someone using this framework would proceed on an ad hoc basis in discussing environmental factors. Most TWVMs use the normal channels of petition for government assistance but lack the power to obtain an adequate response. Part of their problem is the lack of institutionalized procedures and budgeted resources for dealing with a relatively new type of hazard to the community. Hence the TWVM must find alternative leverage points to obtain redress. If these fail the TWVM probably will appeal to bystander publics to gain resources and create greater political pressure. The following observation on movements by Freeman applies also to TWVMs, "Movements that cannot find leverage points are very dependent on the reactions of bystander publics, and it is not uncommon for demonstrations to be used not to affect a particular target directly but to gain sympathy and support from other parties." (pp. 187, 188.)

The above framework of Freeman provides a useful list of concepts for understanding the dynamics of TWVMs and comparing them to other types of social movements. Our major findings from scoring TWVMs on these dimensions are that TWVMs generally have few resources, few constraints, low levels of organization, low opportunities, low opposition, and must appeal to public sympathies to increase support. They are low key, minor social movements which nevertheless often must make a media splash to get some redress of their grievances. Some of these points also emerge when we apply the more processual model of mobilization provided by Gamson, Fireman, and Rytina (1982) to TWVMs below.

Gamson, Fireman, and Rytina emphasize events or encounters, particularly critical incidents, that suddenly change the groups capacity for collective action. These events are seen as turning points in the more continuous and long-term process of mobilization which involve three major steps:

1. Challengers must break the hold over their constituency of the dominant belief systems which legitimize established social and political institutions. Challengers must win people to belief systems that support collective action for change.

2. Challengers must delegitimize the authority that they are challenging in order to mobilize people to their cause. Otherwise normal attitudes of obedience to authority will be a major obstacle to unauthorized collective action.

3. Challengers must create and sustain an organization. They must transform sympathizers into agents and a voluntary association into a managerial structure. (Gamson et al. 1982:6.)

These three steps apply only partially to TWVMs. The first step of breaking the sway of belief systems that support the status quo is common to TWVMs, but the second two steps are not fully applicable. TWVMs usually do not have to compete "with their antagonists, the authorities, for the loyalties of the people whom they are attempting to mobilize" (p. 6). Their potential followers are not particularly loyal to the authorities, but they are very passive. The TWVM does not need to fully delegitimize the authority but does have to shatter victims confidence and trust in authorities in order to make them active. Finally the organizational requirements of TWVMs are much less demanding than organizational requirements for social movements. They also usually face less hostility, less repression and fewer internal control problems. In sum, TWVMs are at the low end of the social movement scale on most of the key dimensions for social movements.

REFERENCES

1. Brown, Michael H.: *Laying Waste.* New York, Pantheon, 1980.
2. California Department of Health Services: Stringfellow Update. February/March, 1989.
3. Citizens's Clearinghouse for Hazardous Wastes: User's Guide to the Citizen's Clearinghouse for Toxic Wastes. Pamphlet, Washington, CCHW, 1987.
4. Coleman, James S.: *Community Conflict* New York, Free Press, 1957.
5. Coleman, James S.: *The Asymetric Society.* Syracuse, Syracuse University Press, 1982.
6. Epstein, Samuel S., Lester O. Brown and Carl Pope: *Hazardous Wastes in America.* San Francisco, Sierra Club Books, 1982.
7. Finsterbusch, Kurt: "The Social Response to Prisoned Environments." Paper presented at the annual meeting of the Southern Sociological Society, Atlanta, April 11, 1987a.
8. Finsterbusch, Kurt: "Typical Scenarios in Twenty Four Toxic Waste Contamination Episodes." Paper presented at the annual meeting of the International Association for Impact Assessment, Barbados, June, 1987b.
9. Freeman, Jo: "Resource Mobilization and Strategy: A Model for Analyzing Social Movement Organization Actions." In Mayer H. Zald and John D. McCarthy (Eds.): *Social Movements.* Cambridge, Mass., Winthrop, 1979, pp. 167–189.
10. Freudenberg, Nicholas: "Citizen Action for Environmental Health: Report on a Survey of Community Organizations." *American Journal of Public Health,* 74, 5, May 1984a, pp. 444–448.
11. Freudenberg, Nicholas: *Not in Our Backyards! Community Action for Health and the Environment.* New York, Monthly Review Press, 1984b.
12. Gamson, William A., Bruce Fireman and Steven Rytina: *Encounters with Unjust Authority.* Homewood, Dorsey, Press, 1982.
13. Gibbs, Lois Marie: *Love Canal: My Story.* Albany, State University of New York Press, 1982.
14. Levine, Adeline Gordon: *Love Canal: Science, Politics, and People.* Lexington, Lexington Books, 1982.
15. Mumpower, Jeryl and Barry F. Anderson: "Causes and Correctives for Errors of Judgment." In Kurt Finsterbusch, et. al. (Eds.): *Social Impact Assessment Methods.* Beverly Hills, Sage, 1983, pp. 241–262.
16. Olson, Mancur: *The Logic of Collective Action: Public Goods and the Theory of Groups.* Cambridge, Harvard University Press, 1965.
17. Schnaiberg, Allan: *The Environment: From Surplus to Scarcity.* New York, Oxford University Press, 1980.
18. Stone, Clarence H.: "Systemic Power in Community Decision Making: A Restatement of Stratification Theory." *American Political Science Review,* 84, 1980, pp. 978–990.
19. U.S. Environmental Protection Agency: *Damages and Threats Caused by Hazardous Materials Sites.* Washington, January, 1980.

20. U.S. Environmental Protection Agency: *Hazardous Wastes Sites: Descriptions of sites on Current National Priorities List.* October, 1984.
21. U.S. Environmental Protection Agency: Personal Communication, March 21, 1989.

Chapter 5

TOXIC WASTE DISPOSAL SITES AS TECHNOLOGICAL DISASTERS[1]

DUANE A. GILL AND J. STEVEN PICOU

INTRODUCTION

Toxic waste disposal sites (TWDS) represent an ominous threat of environmental degradation and technological disaster to our society. Over 125 billion pounds of toxic waste are generated each year by American industry; almost 90 percent of these wastes are improperly disposed. The Environmental Protection Agency (EPA) currently has over 800 waste sites on the national priority list slated for cleanup under the Superfund legislation and has another 20,000 sites under investigation. Perhaps the most staggering figure is the estimated 425,000 waste sites in our country requiring some form of corrective action.

Toxic wastes pose risks to society when individuals, groups, and communities are exposed to excessive amounts of these substances. Exposure to various toxic wastes can cause cancer, birth defects, genetic damage, neurological damage, liver disease, and/or numerous other health problems. The continued production and inadequate disposal of toxic wastes increases the potential of exposure as demonstrated by various communities reporting toxic waste problems. Furthermore, public perceptions regarding the risk of exposure to toxic wastes are heightened as communities and society realize that hazardous waste do not simply degrade into harmless substances, do not remain where stored, and all too often never reach a known or approved disposal site.

The technical problem of properly disposing of toxic waste is as much a sociopolitical issue as it is a technological issue. However, research on the sociopolitical impacts of toxic waste disposal sites has been sparse. As

[1]This research was supported, in part, by the Mississippi Agricultural Forestry Extension Service (Project # MIS 4315) and the University of South Alabama Natural Science Foundation EPSOR program.

a result, there is little information on public perceptions of risks from toxic waste sites and limited awareness of how communities should respond when risks are perceived to be high. Society has difficulty in identifying and agreeing upon the social and health impacts from toxic wastes. Further, establishing responsibility for negative social and health impacts and for cleanup of toxic sites remains elusive. As more toxic waste site situations develop, communities need to prepare for the potential impacts and government agencies need to establish a more comprehensive policy.

This chapter conceptualizes toxic waste sites as actual technological disasters and identifies and reviews the relevant literature on this topic. This conceptual approach provides an established body of disaster research and knowledge which can bring to bear on toxic waste issues. First, we define technological disasters and establish conceptual linkages with toxic waste disposal site situations. This section is followed by a general review of sociological research on the effects of disasters and technological disasters. Next, sociological data and information are summarized on toxic waste sites with an emphasis on Love Canal. Finally, implications for research and policy needs are summarized.

TECHNOLOGICAL DISASTERS

The term "technological disaster" is an emerging concept within the social sciences. An occasion is typically defined as a disaster if it disrupts the social structure's ability to reasonably ensure biological survival, social order, social meanings, motivations, and social interaction. While there is an established research tradition on disaster events with natural causes (e.g., hurricanes, earthquakes, tornadoes, and volcanic eruptions), technological disasters represent a more recent phenomena and are qualitatively different from natural disasters. Following Barton's (1969) collective stress definition of disasters, a technological disaster is:

> ... a situation in which several members of a social system fail to receive the expected conditions of life from the system as a result of unplanned and unfavorable changes in the environment induced directly or indirectly by technological malfunctions (Gill, 1986, pp. 33–34).

Technological disasters involve a *loss of control* over an activity that was once perceived as being controlled whereas natural disasters imply a *lack of control* over processes that are uncontrollable (Baum et al., 1983). The loss of control includes mechanical and structural failures as well as breakdowns in social organizations (Perrow, 1984).

Compared to natural disasters, technological disasters usually lack forewarning, often involve radioactive or other hazardous materials and have long-term impacts. Furthermore, most communities are not socially prepared to respond to a technological disaster. As a result, information flows and expert intervention patterns become inadequate.

Technological disasters have the potential for litigation and raise issues regarding responsibility for clean-up and other socioeconomic costs. A different response to technological disasters may occur within the disaster community. Survivors of technological disasters exhibit more "... anger, hostility, and rage than do victims of natural disasters" (Ahearn, 1981, p. 15; Baum et al., 1983; Couch and Kroll-Smith, 1983). The social psychological impacts are also greater for technological disasters and research suggests that long-term and chronic levels of stress persist among victims of technological disasters and that stress may actually increase over time (Gill, 1986; Picou, 1984).

Toxic waste disposal sites become technological disasters when the impacts of exposure to toxic substances seriously disrupt the host community. These events are perceived to result from a loss of control over a controllable technological process. The perceived threat of exposure debilitates the social structure to the degree that collective stress is experienced.

Toxic waste disposal site disasters are unique in terms of the chronic characteristics displayed by such events (Couch and Kroll-Smith, 1985). Instead of striking quickly and engendering momentary, conspicuous devastation, chronic disasters emerge gradually, sporadically, and may lack certainty as to the nature and degree of devastation. As a result, a definition of the situation gradually develops within the community and events may emerge to create a technological disaster. For example, Fowlkes and Miller (1982) note that the ambiguous nature of the events at Love Canal forced individual families into the unusual and difficult position of developing their own definition concerning the significance of the presence of toxic substances.

Currently, the majority of toxic waste sites are not considered to be disaster areas. Many sites do pose active risks however, and other sites may emerge as disaster situations. These latter instances require preparation of response plans to mitigate negative consequences. Research findings from natural disasters provide a preliminary account of the general patterns of reactions likely to be experienced. The relatively few studies of technological and chronic disasters provide more specific indicators of expected impacts. An understanding of these issues is useful for the development of public policy strategies for more adequately responding to such situations.

REVIEW OF IMPACTS OF NATURAL AND TECHNOLOGICAL DISASTERS

Disasters can be conceptualized into three basic components: disaster consequences, disaster structure, and social characteristics of the disaster area (Gill 1986). Research findings suggest a number of potential consequences for the disaster community. The disaster structure and the social characteristics of the disaster area play a major role in determining the severity of these impacts. This section provides a general overview of the basic findings of disaster research. Data from technological disasters are drawn from the following incidents: Buffalo Creek, WV, 1972 (dam break); Love Canal, NY, 1978–1980 (toxic waste site); Three Mile Island, PA, 1979 (nuclear plant accident); Mississagua, Canada, 1979 (train derailment); and Livingston, LA, 1982 (train derailment).

Disaster Consequences

The distinguishing characteristic of an occasion which qualifies it to be a disaster is the consequences. Disaster impacts include physical destruction, ecological changes, and collective stress. Disaster agents inflict physical impacts such as death and injury, destruction of property, and ecological damage. Collective stress refers to disruptions in the social order which typically result from the physical destruction and danger experienced by victims. Empirical findings suggest different categories of collective stress including: community/organizational, family/small group, social psychological, and individual.

Community/Organizational Stress

Community/organizational stress reflects changes in organizational structures and social relationships within the impact area. Organizational stress occurs when the demands made on the organization exceed the organization's capability to respond. Disasters typically impact organizations such as police and fire departments, hospitals, public utilities, civil defense, and local government. These organizations typically have response plans and experience with natural disasters. Technological disasters, particularly those involving hazardous substances, may present unique situations for which most communities and emergency organizations are not prepared.

Family/Small Group Stress

Family and small group stress are identifiable in terms of changes in behavior patterns, group structures, and social support networks. The family is an important group which mediates disaster impacts on the individual (Bolin, 1982; Bolton, 1979). Studies indicate that in the aftermath of a disaster, families tend to develop greater internal solidarity which is also extended to friends and relatives (Bates et al., 1963; Fritz, 1961). Crawshaw (1963) noted, however, that family conflict can result when members seek to discharge anxiety caused by the disaster. Bolin (1982) notes that victims with heavy losses are more likely to experience a decrease in neighboring than are victims with moderate losses, but, they are also more likely to experience increased visitation from kin. Victims who suffer the loss of a family member or loved one are more likely to be severely disturbed than those who lost only material possessions (Glesser et al., 1981).

Technological disasters provide some unique sources of stress for families. Since technological disasters are relatively unpredictable, families are more likely to be separated at the time of the disaster (Picou 1984). Research indicates that separated families experience greater stress than do intact families (Drabek 1986; Gill 1986; Picou 1984). Furthermore, the uncertainty regarding the impacts from toxic substances may undermine the family's ability to cope with the event, particularly when young children are present. Additional family stress is experienced when families are forced to define a situation as a disaster rather than having the situation clearly defined by its obvious physical consequences (Fowlkes and Miller 1982).

Social Psychological Stress

Social psychological stress refers to unplanned changes in social attitudes and perceptions which occur in order to account for the disaster experience. This type of stress may reflect a discrepancy which emerges between the perceived quality of life and the expected quality of life.

Such changes in attitudes and values may establish a social psychological basis for stress among community members.

Social psychological stress is especially apparent in technological disasters. The literature suggests three types: disruption perception, future risk perception and community satisfaction. Disruption perception refers to long-term feelings of apprehension, danger, threat, disturbance, and uncertainty related to the initial responses to the disaster event. Future risk perception indicates an increased awareness of vulnerability to a variety of hazards (e.g., accidents, chemical/radiation exposure, health-risks, and awareness of potential other disasters). Community satisfaction refers to attitudes and perceptions regarding the quality of social life in the community.

Individual Stress

Stress at the individual level is based on psychological reactions and behavioral responses of victims. Psychological stress may emerge as a "disaster syndrome," dejection, depression, psychosomatic symptoms, and survivor guilt. Behavioral responses indicative of stress include hyperactivity, change in role behavior, and increased consumption of alcohol and other drugs.

The degree of psychological stress is related to the individual's abilities to adapt. Adaptation depends on the appraisal of the stressor, adequate coping techniques, defense mechanisms and prior psychological adaptations. For most individuals, individual stress from natural disasters occurs for a short time period (Barton, 1969; Quarantelli and Dynes 1970; Wallace, 1956) Research from technological disasters, however, indicates more long-term stress (Ahearn, 1981; Houts et al., 1980; Baum et al., 1983; Glesser et al., 1981; Newman, 1976; Titchener and Kapp, 1976; Gill, 1986; Picou, 1984). Since the individual and psychological adaptations to technological disasters are limited relative to natural disasters, individual stress is more likely to occur and persist through time for technological disasters.

Disaster Structure

Disaster structure refers to characteristics of a disaster agent. These factors include the source (i.e., natural or human-made), speed of onset, scope of impact, duration of impact, and controllability. Some disasters are more disruptive than others. Of particular interest are those characterized by: (1) suddenness and uncertainty (Fritz, 1957; Zurcher, 1968); (2) prolonged duration (Fritz, 1957); (3) a broad scope of physical destruction, death and injury (Fritz, 1957); (4) occurrences at night (Lomnitz, 1970) and; (5) exposure to dead and/or seriously injured individuals (Fritz and Marks, 1954).

The empirical evidence from technological disasters also supports an association between the disaster structure and disaster impacts. Findings suggest that groups of individuals living closest to the disaster site are more likely to: (1) experience higher levels of psychological stress (Baum et al., 1983; Houts and Goldhaber, 1981; Houts et al., 1980; Picou, 1984); (2) have a higher perception of health risk (Fowlkes and Miller, 1982; Picou, 1984); (3) evacuate the area by choice (Fowlkes and Miller, 1982; Ziegler et al., 1981); and (4) permanently relocate away from the area (Fowlkes and Miller, 1982). In addition, Glesser et al. (1981), note that Buffalo Creek victims who experienced high levels of economic and personal loss also manifested high levels of psychosocial stress.

Social Characteristics of the Impact Area

Disaster impacts vary according to the social characteristics of the impact community. These characteristics include patterns of interaction, demographic characteristics, economic resources, social attitudes and values, and frames of reference which structure perceptions. Research on disasters indicate several trends regarding social characteristics and disaster impacts. General conditions which have differential consequences within the impact community are resource availability, community integration, and demographic characteristics.

Resource Availability

Resource availability refers to resources within the impact area which aid in coping with the consequences of the disaster. Socioeconomic status is a basic indicator of resources. For example, Bates, Fogleman, Parenton, Pittman, and Tracy (1963) found that victims of Hurricane Audrey with lower socioeconomic status were more likely to suffer negative consequences than victims with higher socioeconomic status. Evacuation behavior is also influenced by socioeconomic status. The poor are more likely to use public shelters and middle income families are more likely to evacuate to private homes or motels (Moore et al., 1963, p. 124). Furthermore, differences in socioeconomic status reflect differences in attitudes and perceptions which, in turn, influence subjective interpretations and styles of coping with the disaster event. These findings support the proposition that victims with greater social and economic resources suffer less negative consequences than victims with less resources.

With regard to technological disasters, there are few findings which directly address the relationship between resource availability and disaster consequences. Glesser, Green, and Winget (1981) found that victims with higher levels of education tended to have less psychological impairment at Buffalo Creek. Fowlkes and Miller (1982) noted similar findings for Love Canal.

Community Integration

A second set of conditions which mediate the consequences of a disaster is community integration. Community integration refers to the social attachments within a community in the form of family, neighborhoods, membership in both informal and formal groups and organizations. Community integration provides social support mechanisms for coping with and interpreting disaster events. For example, Barton (1969) found that normal primary-group ties provided assistance to victims who belong to such groups, while socially isolated individuals often received considerably less support. In addition, Barton (1969, p. 81) noted that there was more panic among those individuals without dependents. Research findings also revealed that persons separated from their families during the disaster tend to experience more negative consequences than do persons who are not separated from family members (Fritz and Marks 1954). Wallace (1957) has noted that the disaster situation may improve individual cooperation and serve to release communal feelings, renew fellowship with others, and provide a rebirth of community and individuals. Although empirical documentation of the consequences of such ties has been sparse (Mileti et al., 1975), literature suggests: the greater the integration of individuals into the community, the less negative will be the social psychological consequences of a disaster.

The literature on technological disasters provides an additional perspective with regard to community integration and family ties. Empirical findings suggest that families which have dependent children are more likely to experience negative social impacts. Houts and Goldhaber (1981) found that families with young dependent children at Three Mile Island were likely to have evacuated the area. Similarly, Fowlkes and Miller (1982) found that families with young dependent children were more likely to permanently relocate away from Love Canal. Studies of the Mississauga and Livingston train derailments produced findings that families with dependent children were more likely to experience higher levels of social psychological and psychological stress (Institute for Environmental Studies 1981; Gill 1986; Picou 1984). Stress among these families was fostered by perception of increased health risks for infants and children.

Demographic Groups

The disaster research literature notes that certain demographic groups are more vulnerable to negative disaster impacts than others. For example, Form and Nosow (1958, p. 103) found that women experience more negative consequences than men. Friedsam (1961; 1962) reported that the elderly were more likely to become physical casualties and experience a greater sense of deprivation than younger adults. Perry and Perry (1959) suggested that younger children experience more psychological stress than adults. These findings imply that levels of social preparedness to cope with disasters varies according to demographic characteristics.

Empirical findings suggest a relationship between demographic characteristics and impacts from technological disasters. For example, women had higher levels of psychopathology at Buffalo Creek than did men (Glesser et al., 1981). With regard to age, adolescent children at Buffalo Creek, Livingston and Mississauga were more prone to psychological disturbances (Glesser et al., 1981; Institute for Environmental Studies, 1981; Newman, 1976; Picou, 1984). Many older residents at Love Canal had a lower perception of risk than did younger adult residents (Fowlkes and Miller, 1982). Similarly, data indicated older TMI residents were less likely to evacuate the area than were younger residents (Houts and Goldhaber 1981).

Summary

This brief review emphasized some of the unique qualities of technological disasters. Since toxic waste sites have characteristics which are similar to technological disasters, these findings can sensitize researchers to potential social impacts. In particular, technological disasters are anomic situations for which few communities, families, and individuals are prepared. Community emergency response organizations may not be familiar with these situations and the uncertainties may be highlighted in families with dependent children. These findings and issues should be considered when investigating communities which are impacted by toxic wastes.

TOXIC WASTE INCIDENTS AND ISSUES

Toxic waste sites pose a variety of problems in addition to those associated with natural and technological disasters. They may evolve as chronic problems and create public perceptions of uncertainty of long-term health effects. Uncertainty, communication problems, and a mistrust of government, industry, and science can create a social milieu of anomie. This section reviews some of these issues as they emerged from Love Canal and present additional issues from other toxic waste cases.

Love Canal

Love Canal exemplified the disastrous impacts of a toxic waste disposal site. Located in Niagara Falls, New York, Love Canal was a dumping site for Hooker Chemical Company in the late 1930s and throughout the 1940s. The dump site was eventually covered over and sold for community development in the 1950s. Although scattered incidents involving "mysterious" substances and unusual illness occurred in the late 1950s through the mid 1970s, it was not until the late 1970s that Love Canal began to be defined as a toxic waste disposal disaster by local residents.

As early as 1976, it was known that Love Canal was an extremely serious problem and in 1978, "... benzine, a known cause of cancer in humans, had been readily detected in the household air up and down the streets" (Brown 1982, p. 301). In August, 1978, a cooperative federal and state program was initiated to purchase the homes of approximately three-hundred families within the immediate vicinity of Love Canal. However, residents remaining in the area surrounding the immediate Love Canal vicinity continued to express apprehension and reservations regarding health risks associated with the migration of toxic chemicals into their homes. In 1980, the relocation program was expanded to include residents in this perimeter. The second relocation was based upon the recognized social-psychological and psychological impacts associated with the uncertainty of the health risks. This was construed as a mental health emergency (Fowlkes and Miller 1982, p. 20). By 1982, 402 of 555 homes in this designated area had been purchased by the federal and state government program.

Brown (1979; 1982), Levine (1982; 1983), and Fowlkes and Miller (1982, 1987) identified many of the problems associated with Love Canal and other toxic waste disposal sites. For example, the toxicity of chemical substances was often unknown. Citizens were usually without reliable information, and public officials attempted to minimize the perception of danger by finding ways to present information without alarming the public. Meanwhile, local and state government, municipal agencies, and corporations argued over liability and responsibility for cleanup. The Love Canal situation resulted in active citizen mobilization over a period of two years which subsequently generated pressure on the Federal government to assist in disaster cleanup and relief (Gibbs 1982).

Levine (1983) identified problems of misinformation and residents' loss of trust in science, government, and officials who managed the situation. By 1980, the situation had deteriorated to the point where organized residents boycotted government sponsored health studies. This breakdown of communication between officials, scientists, and the public was a source of stress in the community. The uncertainty of information led to various perceptions within the community. Not all residents defined the situation as dangerous and as the situation evolved, residents became more polarized on perceptions of risk. Fowlkes and Miller (1987) concluded that the polarization damaged the social fabric of the neighborhood.

Fowlkes and Miller (1982) conducted a quantitative study of Love Canal. They proposed that disasters are socially constructed when there is no "sudden impact" and when there is ambiguity regarding the boundaries of "destruction." These socially constructed definitions of the situation have subsequent powerful and real consequences.

Using the definition of the situation approach, the researchers focused their attention on measuring beliefs and attitudes of Love Canal residents. Data were collected from 63 in-depth interviews of homeowners in the Love Canal area. A stratified random sample was used to select a group of 47 relocated residents and a group of 16 remaining residents. Comparisons between these groups were made regarding perceptions of exposure and health risk and certain demographic and residential characteristics. Three-fourths of those who relocated perceived widespread contamination and an accompanying serious health risk. Over 80 percent of those residents who remained believed the contamination was limited and posed no threat to their health. This difference was related to the residential distance from the site.

Fowlkes and Miller (1982) noted that demographic and social factors were significant when controlling for distance from the site. The sample group of residents who remained in the area were more likely to be older and to have lived in their homes for a longer period of time. More importantly, the majority of these residents had not experienced any "unusual" health problems. These individuals also expressed higher levels of general trust in the government. In contrast, those individuals who decided to relocate were primarily younger adults with dependent children. Many of these individuals had experienced what they perceived to be unusual health problems and tended to distrust the government, particularly its response to the Love Canal crisis. The study also reported that some individuals decided to relocate because they perceived a reduction in community satisfaction resulting from loss of neighbors and physical deterioration in the neighborhood.

In sum, Love Canal was perceived by many residents to be a disastrous situation. Their responses were typical of disaster induced collective stress. Although they were relocated and had their homes purchased by the government, residents from the immediate vicinity lived in a milieu of uncertainty. The uncertainty of exposure amounts, toxicity effects of chemicals, health effects, degree of danger, and latent impacts was exacerbated by unreliable scientific and government information and the inability to fix responsibility for clean-up and restoration.

Other Incidents and Issues

There have been numerous incidents involving toxic wastes in the past two decades. Nader, Brownstein, and Richard (1981) and Epstein, Brown, and Pope (1982) documented several accounts of toxic waste exposure and developed some of the implications for communications and society. For example, Nader and colleagues point out various types of exposure including occupational exposure, air pollutants, water pollutants, chlorination and carcinogens, groundwater contamination and toxins in the food chain. Epstein, Brown, and Pope (1982) identified categories of hazards attributed to toxic wastes including: contamination of groundwater/water supplies, well closures, habitat destruction, human health, soil contamination, fish kills, livestock kills, toxic discharges into public sewers, and damages to crops and wildlife.

Other cases of toxic waste disposal problems have been documented. For example in Woburn, Massachusetts, toxic wastes had been dumped for over 120 years resulting in more than 60 acres of land containing known contamination and three hundred acres which contained " . . . uncertain but suspicious contents" (Ripstein, 1982, p. 46). In the 1970s, several health problems were reported in this working class community of 35,000 residents and subsequent investigation confirmed that Woburn did indeed have a problem. Citizen protest groups formed to voice their concern and their efforts led to the special EPA "superfund" being made available to clean up the Woburn area.

Harris (1983) examined the community history of dumping policies near Memphis, Tennessee. The reactions to potential dangers in this community cut across racial barriers and united citizens into a "task force" to participate in matters of testing and cleanup. While actual migration of toxic substances from disposal sites was not confirmed and no clear health effects were established, public and media perceptions of the risk were indicative of a disaster situation.

Details on other toxic waste disposal sites and incidents further indicate that public perceptions and attitudes are important in defining events as a "disaster." Jasen (1983) contends that chemical waste problems are often ignored because of economic problems in the host community. For example, in Charles City, Iowa, a local veterinary manufacturing company had buried toxic chemicals in an eight and one-half acre area for 25 years. It was later learned that this site was directly above the underground water supply used for most of northeastern Iowa. However, the community did not appear alarmed about potential dangers because economic problems of high unemployment overshadowed potential risks from toxic waste.

Other studies on toxic wastes have focused on accidental and deliberate dumping of toxic wastes into public areas and uncertified dump sites. For example, waste oil applied to the streets of Times Beach, Missouri was discovered to contain dioxin; a dangerous cancer causing chemical. The 2,500 residents of Times Beach were relocated and eight hundred homes were purchased by the federal government at a cost of 36.7 million dollars (Beck et al., 1983, p. 20; Sun, 1983). Researchers have also investigated illegal dumping of wastes on the public highways and other public areas. This practice of "midnight dumping" has been linked to organized crime and poor enforcement of regulations (Beck et al., 1983; Block and Scarpitti, 1985; Epstein et al., 1982).

CONCLUSIONS

An understanding of the community impacts of toxic waste disposal site discovery, or accidental discharge and spillage of toxic wastes, is a serious research issue relevant to a vast number of citizens, citizens groups, chemical producers, waste transporters, and government agencies. Any effective system to reduce the negative impacts of stored toxic wastes on communities must be based on accurate information. More in-depth research on the impacts of toxic waste disposal sites and accidents for the community, family, and individual is needed. Carefully designed scientific studies, which are not viewed by community residents with distrust, can be developed and implemented (Gill and Picou 1989). Such research would not only provide critical information for public policy, but would also provide information which can be utilized in class-action litigation (Picou and Gill 1985).

Many impacts of toxic waste disposal sites remain unidentified. The difficulty identifying and estimating social impacts from waste sites is a direct result of the lack of consensus among scientists, government officials, industry executives, public health officials, and the lay public. This lack of consensus is caused by a failure to resolve the following issues: (1) accurate identification and measurement of toxic substances in an area; (2) determination of the degree of human tolerance to exposure to levels of specific toxic substances (and combinations of substances and exposure time); (3) relationships between community health aberrations and exposure to toxic substances (short-term and long-term); (4) scientific, governmental and public procedures for determining which sites are dangerous (and how to respond to and mitigate the danger); and (5) determination of legal responsibility for short-term and long-term impacts. The resolution of these issues requires awareness of the nature of toxic waste storage problems in our society. These problems are compounded by the unique features associated with technological disasters. Hopefully, this chapter has increased our awareness of the potential impacts of toxic waste sites on the community. Such awareness should lead to a consideration of strategies and policies for initially addressing the issues identified in this chapter.

REFERENCES

Ahearn, Frederick L.: Man-made and natural disasters: a comparison of psychological impacts. Paper presented at the annual meeting of the *American Public Health Association,* Los Angeles, 1981.

Barton, Allen H.: *Communities in Disaster: A Sociological Analysis of Collective Stress Situations.* Garden City: Doubleday, 1969.

Bates, Frederick L., Fogleman, Charles W., Parenton, Vernon J., Pittman, Robert H., and Tracy, George S.: *The Social and Psychological Consequences of a Natural Disaster: A Longitudinal Study of Hurricane Audrey.* National Research Council Disaster Study #18. Washington: National Academy of Sciences, 1963.

Baum, Andrew, Fleming, Raymond, and Singer, J.E.: "Coping with victimization by technological disaster." *Journal of Social Issues,* 39(2):117–138, 1983.

Beck, Melinda, Junkin, D., Taylor, J., Mandel, M., Argest, S., and Achiron, M.: "The toxic waste crisis." *Newsweek,* 101(10):20–24, 1983.

Block, Alan A., and Scarpitti, Frank R.: *Poisoning for Profit: The Mafia and Toxic Waste in America.* New York: William Morrow, 1985.

Bolin, Robert C.: *Long-Term Family Recovery from Disaster.* Boulder: University of Colorado, Institute for Behavioral Science, 1982.

Bolton, P.: *Family Recovery Following a Natural Disaster: The Case of Managua, Nicaragua.* Ph.D. dissertation, Boulder: University of Colorado, 1979.

Brown, Michael: *Laying Waste: Love Canal and the Poisoning of America.* New York: Random House, 1979.

Brown, Michael: Love Canal and the poisoning of America. In Skolnick, Jerome, and Currie, Elliot: *Crisis in American Institutions.* Boston, MA: Little Brown, 1982, pp. 297–315.

Couch, Stephen R., and Kroll-Smith, J. Stephen: The chronic technical disaster: toward a social scientific perspective. *Social Science Quarterly,* 66(3):564–575, 1985.

Crawshaw, Ralph: Reactions to a disaster. *Archives of General Psychiatry,* 9(August): 157–162, 1963.

Drabek, Thomas E.: *Human Systems Responses to Disaster: An Inventory of Sociological Findings.* New York: Springer-Verlag, 1986.

Epstein, Samuel S., Brown, Lester O., and Pope, Carl: *Hazardous Waste in America.* San Francisco: Sierra Club, 1982.

Form, William H., and Nosow, Sigmund: *Community and Disaster.* New York: Harper, 1958.

Fowlkes, Martha R., and Miller, Patricia Y.: *Love Canal: The Social Construction of Disaster.* Washington: Federal Emergency Management Agency, 1982. Fowlkes, Martha R., and Miller, Patricia Y.: Chemicals and community at Love Canal. In Johnson, Brander B., and Couello, Vincent T. (Eds.): *The Social and Cultural Construction of Risk.* Boston: D. Reidel, 1987, pp. 55–78.

Friedsam, Hiram J.: Reactions of older persons to disaster-caused losses: an hypothesis of relative deprivation. *Gerontologist,* 1(1):34–37, 1961.

Friedsam, Hiram J.: Older persons in disaster. In Baker, George W., and Chapman, Dwight W. (Eds.): *Man and Society in Disaster.* New York: Basic, 1962, pp. 151–184.

Fritz, Charles E.: Disasters compared in six American communities. *Human Organization,* 16(Summer):6–9, 1957.

Fritz, Charles E.: Disasters. In Merton, Robert K., and Nisbet, Robert A. (Eds.): *Contemporary Social Problems.* New York: Harcourt, 1961, pp. 651–694.

Fritz, Charles E., and Marks, Eli S.: The NORC studies of human behavior in disaster. *Journal of Social Issues,* 10(3):26–41, 1954.

Gibbs, Lois M.: *Love Canal: My Story.* New York: Grove, 1982.

Gill, Duane A.: *A Disaster Impact Assessment Model: An Empirical Study of a Technological Disaster.* Ph.D. dissertation, College Station: Texas A&M University, 1986.

Gill, Duane A., and Picou, J. Steven: A research design for disaster impact assessment. Paper submitted to *International Journal of Disaster and Mass Emergencies,* 1989.

Glesser, Goldine C., Green, Bonnie L., and Wignet, Carolyn C.: *Prolonged Psychological Effects of Disaster: A Study of Buffalo Creek.* New York: Academic, 1981.

Harris, Jeffrey S.: Toxic waste uproar: a community history. *Journal of Public Health Policy,* 4(2):181–201, 1983.

Houts, Peter S., and Goldhaber, Marilyn D.: Psychological and social effects on the population surrounding Three Mile Island after the nuclear accident on March 28, 1979. In Majumdar, S. (Ed.): *Energy, Environment and the Economy.* Harrisburg: Pennsylvania Academy of Sciences, 1981, pp. 151–164.

Houts, Peter S., Miller, R.W., Ham, K.S., and Tokuhata, G.K.: Extent and duration of psychological distress of persons in the vicinity of Three Mile Island. *Proceedings of the Pennsylvania Academy of Science,* 54:22–28, 1980.

Institute for Environmental Studies (IES): *The Mississippi Evacuation Final Report.* Toronto: University of Toronto, 1981.

Jasen, Georgette: "Iowa town learns to live with chemical waste." *Wall Street Journal,* 201(April 27):60, 1983.

Levine, Adeline: *Love Canal: Science, Politics and People.* Lexington: D. C. Heath, 1983.

Levine, Adeline: Psychological impact of toxic chemical waste dumps. *Environmental Health Perspectives,* 48:15–17, 1982.

Lomnitz, Cinna: Casualties and behavior of populations during earthquakes. *Bulletin of Seismological Society of America,* 60(August):1309–1313, 1970.

Mileti, Dennis S., Drabeck, Thomas E., and Haas, J. Eugene: *Human Systems in Extreme Environments: A Sociological Perspective.* Boulder: University of Colorado, Institute of Behavioral Science, 1975.

Moore, Harry Estill, Bates, Frederick L., Layman, Marvin V., and Parenton, Vernon J.: *Before the Wind: A Study of Response to Hurricane Carla.* National Research Council Disaster Study #19. Washington: National Academy of Sciences, 1963.

Nader, Ralph, Brownstein, Ronald, and Richard, John: *Who's Poisoning America? Corporate Polluters and their Victims in the Chemical Age.* San Francisco: Sierra Club, 1981.

Newman, Janet C.: Children of disaster: clinical observations at Buffalo Creek. *American Journal of Psychiatry,* 13(3):306–312, 1976.

Perrow, Charles: *Normal Accidents: Living with High Risk Technologies.* New York: Basic, 1984.

Perry, Helen S., and Perry, Stewart E.: *The School House Disasters.* National Research Council Disaster Study #11. Washington: National Academy of Sciences, 1959.

Picou, Steven: *Ecological, Physical, Economic, Sociological and Psychological Assessment of The Illinois Central Gulf Train Derailment. Vol. 5: Sociological Assessment.* Baton Rouge: Gulf South Research Institute, 1984.

Picou, J. Steven, and Gill, Duane A.: Social impact assessment as a legal response to technological disasters: the case for latent social injury compensation. Paper presented at the annual meeting of the *Southern Sociological Society,* Charlotte, 1985.

Quarantelli, Enrico L., and Dynes, Russell R.: Property norms and looting: their patterns in community crisis. *Phylon,* 31(Summer): 168–182, 1970.

Ripstein, Diane: Chemical waste beneath Massachusetts town. *Business and Society Review,* 41:46–47, 1982.

Sun, Marjorie: Missouri's costly dioxin lesson. *Science,* 219:367–369, 1983.

Titchener, James L., and Kapp, Frederick T.: Family and Character change at Buffalo Creek. *American Journal of Psychiatry,* 13(3):295–299, 1976.

Wallace, Anthony F.C.: *Tornado in Worcester.* National Research Council Disaster Study #3. Washington: National Academy of Sciences, 1956.

Wallace, Anthony F.C.: Mazeway disintegration: the individual's perception of sociocultural disorganization. *Human Organization,* 16:23–27, 1957.

Zeigler, Daniel J., Brunn, Stanley D., and Johnson, James J. Jr.: Evacuation from a nuclear technological disaster. *The Geographical Review,* 71(1):1–16, 1981.

Zurcher, Louis A.: Social-psychological functions of ephemeral roles: a disaster work crew. *Human Organization,* 27(Winter):281–297, 1968.

Chapter 6

CARGOISM AND TECHNOLOGY AND THE RELATIONSHIP OF THESE CONCEPTS TO IMPORTANT ISSUES SUCH AS TOXIC WASTE DISPOSAL SITING

WILLIAM R. CATTON, JR.

By temperament and upbringing I am an optimist. As an optimist my natural inclination would be to proclaim that, given enough effort and good will, our environmental problems are going to be solved. I would want to say that our grandchildren will live happily in a clean, safe, and lovely world. I would acknowledge that today we have worries about serious health hazards from toxic waste dumps and from air and water pollution, but I would want to insist that "our glass is half full, not half empty."

Despite my optimistic temperament, however, that cannot be my message. Instead, what the evidence says to me is that we are not going to give our descendants a clean and lovely world; their anxieties are not going to be milder than our own. The hazards in their lives will not be less than the ones we today keep trying to deny or downplay. Instead I have to say that in my judgment the words at the beginning of the Global 2000 Report were factually correct: notwithstanding conditional "ifs" about continuation of present trends, a dozen years hence the world *will* "be more crowded, more polluted, less stable ecologically, and more vulnerable to disruption" (Council on Environmental Quality and Department of State 1980:1).

The first strange word in the odd title of my chapter has a lot to do with the reason for this being our destiny. But before defining my key terms, I want to tell how certain experiences have led me to this somber view. And it is appropriate that I should confess some ambivalence.

If I am by nature optimistic then, of course, I would hope the concepts, the evidence, and the reasoning that seem to point to a dire destiny are flawed. But wisdom insists that even unwelcome knowledge is preferable to ignorance. If mankind is walking a tightrope where missteps could be fatal, it is better to have light than darkness. So I consider myself lucky to have had these enlightening experiences.

I was lucky, for example, in the fact that my wife's mother happens to have reached a ripe age where her impaired eyesight and hearing loss required that she give up driving a car, for this meant that in her apartment during one of our visits there happened to be a like-new road atlas that was several years older than the dog-eared one in our car. Her obsolescent atlas made one of my enlightening experiences happen.

Travel is said to be educational. Some trips teach us more than we might be expecting. I am going to explain how some valuable insights became available to me on three trips. Two of them were in the United States, one by automobile and one by air. The third was overseas, an extended sojourn. What we learn from our travels depends on what we have in mind before and after we take a trip. I am going to convey some things I happen to have learned from sandwiching these trips between some fortuitously helpful reading which enhanced the significance of a traveler's observations.

The Dioxin Shocker

The most recent learning excursion I want to describe happened because of my mother-in-law's outdated road atlas. Between her residence in Indiana and ours in the state of Washington, we often pass through Missouri. I had read about the Environmental Protection Agency (EPA) having bought out a whole community in Missouri after that town became irreparably contaminated with the extremely toxic chemical dioxin. I had not known just where in Missouri this town was located. It no longer appears on recently published road maps. But on my wife's mother's pre-1983 map there it was, on an insert showing the St. Louis metropolitan region. It was just off Interstate 44, about 25 miles out from the city. So we decided to route ourselves that way at the close of our Indiana visit, as I hoped to shoot two or three Kodachrome slides that might someday prove useful in a lecture on environmental problems.

All road signs along the freeway that used to indicate the exit for Times Beach had long since been removed, so we had to make local inquiry to find the place. When we did learn where to turn off the Interstate we soon encountered a barricade blocking the two lane road into the ghost town. There were huge warning signs telling us the road was closed, access was restricted, that this was a "Hazardous Waste Site" with "Dioxin Contamination." Just to the left of the barricaded road was a parking lot in front of an abandoned and deteriorating yellowish building that used to be a rural night spot or restaurant, and at the edge of the lot was a trailer office manned by a uniformed and armed guard. When I talked with him, he proved unable to authorize even momentary entry into Times Beach by a camera-toting tourist, but he gave me a telephone number to call.

I went to a pay phone. The number turned out to be a residence in the next town. It belonged to the last mayor of Times Beach. Her name was Marilyn Leistner. Because I identified myself over the phone as an out-of-state professor specializing in environmental sociology, and because a substantial downpour of rain a few hours earlier had settled the hazardous dust, she offered to come in her car and give me and my wife an escorted tour of the dead town. This was more than I had been expecting. It proved to be a very moving experience, for which I was ill-prepared. As she pointed things out to us during an hour's low-gear crawl along the deserted streets of a town formerly inhabited by some two thousand to twenty-five hundred people, most of whose 800 homes were still standing, all vacant and silent, I earnestly wished I had brought a microcassette recorder with me. Unaided, I could not possibly retain all she was telling us about what maladies had befallen the residents of one house after another. Having expected only to take two or three quick photographs on a hurried and unguided visit, I had not even brought a notebook. Probably I could not have scribbled adequate notes at a pace commensurate with her telling anyway. I did have a fresh roll of film in my camera, and through closed windows of her air-conditioned car I shot all 24 exposures in the course of this tour. I remember some of her phrases: "seizure disorder" here, "cancer of the kidney" there, "stomach and bladder problems," "skin rash." That last phrase recurred in almost every block. Among her own four grown children, one is hyperthyroid and had had cervical cancer. Suicidal behavior had also occurred in response to the town's chemical woes.

Before starting this tour, Mayor Leistner had taken my wife and me into that trailer-office of the security guard, where we both had to sign the following "General Release of Liability," a document with many phrases symptomatic of humanity's future:

> I (We), the undersigned, desire to enter the Restricted Area which is comprised of all real estate within the corporate limits of the former City of Times Beach, Missouri.
>
> I (We) understand that said Restricted Area is contaminated with dioxin and other hazardous chemicals and that in addition to such contamination, there exist various hazardous conditions in said area including but not limited to those hazards caused by the deterioration of the buildings, roads and other facilities in said Restricted Area.
>
> I (We) have been informed that entering said Restricted Area without protective suits and other devices is not recommended by the Center for Disease Control. . . .
>
> For and in consideration of the right to enter said Restricted Area, I (we) do hereby release and hold harmless Marilyn Leistner, Trustee of the City of Times Beach, Missouri, and said Trustee's agents, employees, successors and assigns from any and all claims or demands which I (we) may assert as the result of any injury, disease, damage, loss or claim of whatsoever nature which may arise as the result of my (our) presence in said Restricted Area.
>
> I (We) understand that I (we) am (are) assuming all risks which may arise as a result of my (our) presence in the Restricted Area formerly known as the City of Times Beach, Missouri.

The dioxin that shut down this town came into being as a by-product of the manufacture of an antiseptic, hexachlorophene, in a plant at Verona, Missouri, leased by Northeastern Pharmaceutical & Chemical Company (Garmon, 1983). Early in the 1970s the market for hexachlorophene collapsed, and the Independent Petrochemical Corporation of St. Louis was hired to dispose of the dioxin-containing sludge that had accumulated in the process of purifying several years' production of the antiseptic. An enterprising but chemically uninformed subcontractor combined this sludge with waste oil and sprayed some 2,000 gallons of the dust-settling mixture on a horse arena—in whose vicinity birds, rodents, and even horses soon began dying; people, too, began coming down with various illnesses.

From 1972 to 1976, to hold down the dust on its miles of unpaved streets, Times Beach had them regularly sprayed—by the same salvage-oil subcontractor who had elsewhere done that horse arena job (Gough, 1986:121–136). Again he used the sludge-containing mixture. Health complaints in Times Beach over the next several years led to investigation and soil sampling by the EPA. Some efforts to remove the contaminated surfaces of the streets were made, but all hope of restoring the town to livability vanished when a winter flood of the Meramec River spread the contaminated soil from the roads into all the homes on the flood plain (Ozinga, 1985:8–11; Burford, 1986:188). A reluctant EPA took the only recourse, buying out the whole community and terminating its existence as a town.

Some of the pictures I took showed five-digit dollar prices spray-painted in large numerals on the fronts of abandoned houses. Ex-mayor Leistner explained that these were a response of irate citizens to the initial "fair market value" offerings by the EPA. Prices for Times Beach homes were held down by the town's flood plain location. Because of this natural hazard, amounts first offered by EPA would not have sufficed to buy comparable housing in other nearby towns, so the evacuees angrily painted these low values on their homes and invited St. Louis television news camera coverage, thus embarrassing EPA into raising the bids somewhat. According to Leistner, however, in this among many other ways one of the ominous effects of the whole frightful experience was the townspeople's "complete loss of faith in the government."

A year afterward it does seem that her tightly closed car insulated me and my wife from the chemical hazard, but it could not protect us from being sickened even now by memory of something she showed us and described to us. A sign had been put up on the side of a motel-like apartment building that said in huge letters, "There's HOPE Because JESUS IS LORD!" Ex-mayor Leistner saw that this religious expression of ecological naivete had struck me as at best ironic, and this prompted her to tell us of a minister in the area who perceived vacant structures in this horror-stricken community as a heaven-sent solution to an urban social problem. He conceived the idea of housing homeless men from downtown St. Louis in the houses being abandoned in Times Beach. He began bringing the men out to select dwellings even while these were just in process of being sold to EPA. Former residents vacating them, she said, had told her this felt like being watched on their death beds by circling vultures.

The "Clean-Up" Notion

I wish it were possible that everyone who has the slightest doubt as to the seriousness of mankind's environmental predicament could have the experience I had in visiting Times Beach. For me, the demise of this innocent, working class town epitomizes the dread made endemic by our era. But since nearly all Americans have never lived in Times Beach, most can shrug off the death of this one small town as no indication that hazardous waste problems are anything but soluble. Ubiquitous contaminated sites, we may persist in wishfully supposing, *can* be "cleaned up." But that phrase masks reality. Our use of it reflects our thoughtlessness. The same shallowness of thought was implicit in the cover picture on an issue of Newsweek back in June 1972 proclaiming "The Environmental Crisis '72." Under bold red letters spelling the words "THE BIG CLEANUP" there was a picture of a dirty planet Earth with a clean swath across it (including all of the continental United States) behind a wet sponge being wiped along by a cosmic hand. The notion of "cleaning up" our polluted planet is naive because it neglects a principle some of us may vaguely recall being exposed to in high school—the idea that matter cannot be destroyed but only moved, reshaped, or recombined.[1]

In Missouri, when some of the dioxin contaminated soil was removed from sites where it had already caused much grief, it had to be *put somewhere else.* In at least one instance some of it was trucked into a town where surface depressions in a residential area needed fill material. This event made it obvious that the grief had just been relocated, not ended (Ozinga 1985:9). In years to come it will be repeatedly seen that "cleaning up" one part of our densely inhabited planet means putting contaminants into some other part of it. Manufacturing processes will always create unwanted by-products along with desired products; increasingly our only escape from the "by-products" will be to forgo the "products." We are learning that lesson only with glacial reluctance.

The Superstition of Our Time

Now, before telling what I learned from those other travel experiences I alluded to earlier, it is time to define the strange term "Cargoism." Elsewhere I have used this word to mean a superstitious "faith that technological progress will stave off major institutional change even in a post-exuberant world" (Catton 1980:272). By a "post-exuberant" world I mean a world that has already lived through a period of exuberant expansion into previously vacant niches and is now increasingly squeezed by carrying capacity limits. The word "cargoism" was coined to liken our own inordinate faith in technological panaceas to the "cargo cults" of people in the islands of Melanesia. Experiencing "relative deprivation" as their primitive ways came into comparison with the material wealth they observed flowing from unseen overseas sources to the dominant European minority living among them, and seeing no expenditure of labor by the Europeans to produce this wealth, the Melanesians developed belief systems which embodied the supposition that they too might ultimately receive similar cargo (from a supposed spiritual source, without otherwise altering their existing culture) if only they made ritual preparation for its delivery.

The institutions of the United States were largely formed in what can well be called an "age of exuberance," when opportunities for growth and progress were a major fact of life (Catton and Dunlap, 1980:27–31). Now, when our ecological and demographic circumstances are so very different from those that shaped our values, our beliefs, and our traditions, there is resistance to the pressure from those changed circumstances to change our institutions. A major part of that resistance takes the form of an expectation that further technological progress will solve any problem arising from the changed situation, thereby obviating the need for unwanted institutional change. Our excessive faith in future technology is as naive as Melanesian faith in magic. The expectation that ways will be found to "clean up" toxic waste sites is a manifestation of such faith. Our neglect of the principle of conservation of matter is one frequent indication that the faith is naive (albeit somewhat less so than the inadvertently cruel attempt by that ecologically unsophisticated clergyman to house the homeless in houses made vacant by dioxin contamination).

Why are we faced today with such issues as where and how to dispose of toxic wastes? This planet is no smaller than it was when humans first began to generate wastes, but we are vastly more numerous than were those remote ancestors. Moreover, with technological progress we have achieved greatly enhanced waste-generating powers. So there is more bad stuff than ever before to be disposed of somewhere, and the somewheres are already in use.

To add to the confusion, we live in a time when economists, ecologists, politicians, journalists, even clergymen and philosophers, and many others persist in debating whether there are limits to human progress and growth—or the derivative question of when we will confront (or whether we have already begun to encounter) those limits and their effects. For each study someone cites to demonstrate ecological limits, someone else offers a pro-growth rationale and cites a counterstudy or an optimistic essay that purports to disprove limits (cf., e.g., Meadows et al., 1972; Cole et al., 1973; Council on Environmental Quality and Department of State, 1980; Simon and Kahn, 1984). The issue defies resolution because the opposing sides proceed from irreconcilable "worldviews" or "paradigms." So they "talk past each other," whether the clash is viewed as "optimists versus pessimists" (Luten, 1980), as "cornucopians versus catastrophists" (Cotgrove, 1982), as "expansionists versus limitationists" (Russell, 1979), or as "exemptionalists versus ecologists" (Catton and Dunlap, 1980). The two sides differ in the way they interpret facts, but they also differ, perhaps more importantly, in what facts they deem relevant (Dunlap, 1983).

Distinguishing Three Uses of Environments

One reason for the confusion is a common failure to recognize the fact, or the significance of the fact, that all living creatures make three different kinds of use of their environment. For human societies, as truly as for other life forms from amoeba to whale, the biophysical environment has to function as:

1. "dwelling place," or "living space";
2. "supply depot," or source of "sustenance materials";
3. "waste disposal site," or repository for the material products of living.

In short, the environment has to serve as a place *in which* to carry on activities, a place *from which* to make withdrawals of material goods, and a place *into which* the ecological end-products we call "waste" (which have to go somewhere) can be put.[2]

Efforts to solve problems pertaining to one of these functions can, in an increasingly overloaded world, create or intensify problems connected with one or both of the other two functions. The complex interrelations among the three functions on a crowded planet reflect fundamental ecological principles of interrelatedness, feedback, and unintended consequences (Commoner, 1971:33–46; Hardin, 1968:38; Merton, 1936). Before our world entered the postexuberant age, there were plenty of opportunities for segregating the three functions in separate portions of the environment. Segregated, they need not interfere with each other. The "into which" sites could formerly be kept away from the "in which" places and the "from which" sources. But segregation of functions has become less and less feasible in our time, though perhaps more and more essential.

Learning from Sheep

Let us now turn to another travel-induced insight. At the time that the dioxin sludge in Missouri was being mixed with used oil and was about to be sprayed on dusty arenas and roads, I was living far away in New

Zealand, where a land area equal to one-and-a-half Missouris was inhabited by only three-fifths as many people (about three million). In New Zealand there were also, however, some sixty million sheep. It is a beautiful country, and we explored it quite thoroughly and enthusiastically during three years of our family's residence there. Almost everywhere we drove, and many of the places we hiked, there were sheep in view most of the time. One day it occurred to me that if the people were as numerous as those sheep, New Zealand would be a very crowded country. However ubiquitous the sheep were, they seemed not to make it crowded. This set me to thinking, for the average adult sheep is not that much smaller than the average adult human.

When I observed a sheep it was often from the window of my moving car; when a sheep saw me it was almost always by momentarily raising its head from grazing. I might end the day two or three hundred miles from where I'd had breakfast; the sheep might consume its final blade of grass at twilight less than a hundred feet from where it chewed off the first one that morning.

Suddenly, I was acutely conscious of the fact that technology made me a consumer of more space than is needed by an animal without technology. Further, my activities required "sustenance materials" (extracted from my environment) not only to put into my own stomach but also to feed my car's fuel tank, and as a result of my actions CO_2 and other wastes would be inflicted upon my surroundings not only by my own lungs and bodily metabolism but also by my car's "metabolism" of petrol in its internal combustion engine. Exhaust gases emitted from its tailpipe "nostril," unlike the spent air exhaled from my own nostrils, would put into the environment lead particles, hydrocarbons, oxides of nitrogen, etc.

I saw both the human and sheep inhabitants of New Zealand as products of an evolutionary process, of course. Reflecting further on evolution, I began to reconsider the sociologically orthodox views I previously held about differences between biological and cultural evolution. I began to see the latter as a natural extension of the former. New Zealand has so many sheep partly because their wool can be shorn, spun, and woven or knitted into garments to be worn by humans. European settlers of New Zealand introduced the sheep to those islands. I began to see woolen sweaters as "prosthetic devices." Better than the flax and feather garments of the New Zealand Maori, woolen "prostheses" can "restore" to the human body the power to retain as much of its metabolism-generated heat as the sheep's body retains. This is a power of heat retention which the human body might have had without technological assistance were we endowed naturally with a woolly covering as is the sheep—had protohumans not originated in a tropical environment. Cultural evolution provides, among other things, "prosthetic" substitutes for organs or body parts we were not given by our biological heritage.

Years earlier I had been impressed (in Seattle) with devices like the artificial kidney machine, which can perform externally to the human body the hemodialysis no longer being done by a patient's failed kidneys. This had caused me to tell students in my introductory sociology classes to think of material culture as a set of "detachable organs." I regarded the glasses I wore as detachable improvements upon my own myopic eyes. I taught my students to regard libraries and computers as detachable extensions of their brains which enabled human heads to remain small enough to pass through the birth canal even while we could evolve (culturally, technologically) the potential to justify our species name, Homo *sapiens.*

Living for a while in a small country with three times the U.S. sheep population and only 1/70 the U.S. human population enabled me to come to grips with what is most distinctive about our species: we have evolved the ability to evolve prosthetic organs to meet new needs instead of awaiting biological evolution of fleshy organs. Now I saw the jet aircraft that had taken me and my family to New Zealand as a set of prosthetic wings and flight muscles enabling us to live like migratory bird species, shifting the locale of our "in which," our "from which," and our "into which" uses of environment. But I had also to contrast jet exhaust with preprosthetic human exhalation.

On an internal jet flight in New Zealand, en route from my South Island home to a lecturing engagement in the North Island, I had occasion to read a small book I had recently acquired. It was one of the early contributions to environmental sociology, and it was written by another American who had taught for three years in a New Zealand university. Two sentences I read on that plane made a deep and lasting impression: "In order to survive, human societies must exploit their surrounding ecosystem. Under some conditions such exploitation becomes so extreme as to destroy the survival base of the society" (Burch, 1971:49). Pondering those sentences, I came to see that we humans are not exempt from the ecological constraints that apply to all other living things, although, as I have shown elsewhere (Catton, 1980:238–2 0, Catton and Dunlap, 1980), sociologists have shared with popular culture the supposition that our humanness *does* afford exemption. Every molecule we take from our environment and transform—either in our bodies or in our external apparatus—in the course of living has to be put back into our environment, however toxic may be the form into which our use has transformed it. The more "from-whiching" we do, with our advanced technology, the more "into-whiching" has to follow. That is why sixty million human beings, especially modern industrial-way-of-living humans, would crowd New Zealand's 103,000 square miles so much more than did those sixty million contentedly grazing sheep.

Two Dimensions

I had taken my family to New Zealand partly to recover the advantages of an environment with low population density. Much of New Zealand's South Island in the 1970s resembled greatly the mountain country around Washington's Puget Sound in about 1910, before the burgeoning of metropolitan population in Tacoma-Seattle-Everett following World War II. But while living in New Zealand I began to shift my concern to another dimension of the pressure human societies exert upon their environment. Demographic growth is one dimension; technological growth is another. The greater the technology, the larger the per capita impact (and, in general, the larger the per capita appetite for resources). In Table 6-1, if we scan the waste and population columns we can easily see that they are correlated; generally the more populous states produce more waste. (Since r = .76, slightly more than half the variance in the hazardous waste figures is "explained" by state population variance.) But there remains conspicuous variation among the "Tonnes per capita" ratios; this clearly indicates there is another factor in waste production besides number of people.

By the time extended family ties drew us back to North America, I had come to recognize how much more serious are the pressures exerted upon an environment by each million industrialized people as compared with each million preindustrial people. I should have known this already from previous reading (e.g., Day and Day, 1964), but I had remained attached to the customary view that bigger and better technology ("progress") must enlarge an environment's carrying capacity. Now I had come to realize it could also enlarge the load imposed by a given number of inhabitants.

Years earlier I had been strongly influenced by reading Fred Cottrell's seemingly ahead-of-its-time *Energy and Society* (1955). In that book, Cottrell clearly showed how certain major differences among human societies

Table 6-1.
Estimated Generation of Industrial Hazardous Waste, 1983, by State—Compared with State Populations

	Thousand tonnes of industrial hazardous waste	*Population (1000s)*	*Tonnes per capita*		*Thousand tonnes of industrial hazardous waste*	*Population (1000s)*	*Tonnes per capita*
Alabama	6,547	3,893	1.68	Montana	662	787	0.84
Alaska	52	402	0.13	Nebraska	739	1,570	0.47
Arizona	642	2,718	0.24	Nevada	379	801	0.47
Arkansas	3,729	2,286	1.63	New Hampshire	431	921	0.47
California	17,284	23,668	0.73	New Jersey	12,948	7,365	1.76
Colorado	1,902	2,890	0.66	New Mexico	619	1,303	0.48
Connecticut	4,238	3,108	1.36	New York	9,876	17,558	0.56
Delaware	894	594	1.51	No. Carolina	3,954	5,881	0.67
Florida	2,981	9,747	0.31	No. Dakota	269	653	0.41
Georgia	3,338	5,463	0.61	Ohio	19,692	10,798	1.82
Hawaii	202	965	0.21	Oklahoma	2,673	3,025	0.88
Idaho	1,160	944	1.23	Oregon	969	2,633	0.37
Illinois	14,810	11,427	1.30	Pennsylvania	18,260	11,865	1.54
Indiana	10,189	5,490	1.86	Rhode Island	1,745	947	1.84
Iowa	1,774	2,914	0.61	So. Carolina	3,669	3,123	1.17
Kansas	2,564	2,364	1.08	So. Dakota	159	691	0.23
Kentucky	4,647	3,660	1.27	Tennessee	12,159	4,591	2.65
Louisiana	13,801	4,206	3.28	Texas	34,866	14,228	2.45
Maine	337	1,125	0.30	Utah	1,139	1,461	0.78
Maryland	2,989	4,215	0.71	Vermont	226	511	0.44
Massachusetts	4,536	5,737	0.79	Virginia	4,038	5,347	0.76
Michigan	12,399	9,262	1.34	Washington	5,523	4,132	1.34
Minnesota	2,212	4,076	0.54	W. Virginia	5,642	1,950	2.89
Mississippi	1,816	2,521	0.72	Wisconsin	3,297	4,706	0.70
Missouri	6,046	4,917	1.23	Wyoming	572	470	1.22

Sources: Hazardous Waste data, The Conservation Foundation, *Groundwater Protection.* Washington, DC: The Conservation Foundation, 1987, p. 120.

Population data, *The World Almanac and Book of Facts 1986,* p. 261

can be accounted for by the energy sources available to them and the characteristics of the energy converters they use. He defined the important concept "surplus energy" as "the energy available to man in excess of that expended to make energy available." For example, the caloric content of a bag of grain minus the energy spent in producing it would constitute the surplus energy value of that grain. Cottrell showed that when the per capita surplus is low, not only must life be comparatively simple but social change has to be exceedingly slow. Nearly all the energy available (through human metabolism, without much exosomatic supplement) goes into just maintaining life—i.e., obtaining sustenance, or replacing the energy used. But when converters and energy sources are on hand that yield a considerable per capita surplus, a human society becomes capable of elaborate and diverse activity and can escalate social change. We who live in the latter type of society have learned to take that condition so much for granted we may be as oblivious to its prerequisites as fish supposedly are to water.

But Cottrell's distinction between "low energy societies" and "high energy societies" had impressed me, because I happened to read his book during a transcontinental flight to a job interview. Our second son had just been born and I had to cross the U.S. while my wife and new baby were in the hospital. We were then living in a city where we had few close acquaintances. We knew nobody there well enough to leave our 23-month-old firstborn son with during my absence. So I had made a long distance phone call to a cousin who lived in a state near where I was seeking employment, took the toddler with me on the plane, and somehow managed to get some reading done. Reading about "high energy societies" while taking a youngster 2,500 miles overnight to a baby sitter gave the concept special and unforgettable salience.

Still, it took three years of living amongst sixty million sheep to bring me to the realization that high energy societies may be at greater risk than low energy societies that their exploitation of "their surrounding ecosystem . . . becomes so extreme as to destroy the survival base of the society."

Modern industry makes mankind more thoroughly a prosthetic species than ever before,[3] but it is by no means the sole source of environmental pollution; agriculture also pollutes, especially modern agriculture. Anyone who has come within sniffing distance of a cattle feed lot knows how concentrated the "into which" use of an environment can become, and how repulsive can be the result. A feed lot is fundamentally like a city, populated far beyond the carrying capacity of the territory it encloses. The city, as truly as the feed lot, requires both a "from which" and an "into which" hinterland. In the feed lot instance, the economic interest of the meat industry ensures that it will amply provide the imports of sustenance materials required by the livestock, but there is much less urgent concern (among the humans managing the system) with exporting the wastes, so they tend to accumulate on site.

Total wastes from 48.6 million cattle, 57 million hogs, 20.9 million sheep, and 3,000 million fowl in the United States are equivalent to the bodily wastes from two billion humans (CEK, 1970). So, quite apart from industry, the agrarian component of the American nation suffices to make our effective population load equivalent to some 2.25 billion people, not just the 1/4 billion who actually belong to our species and get counted in the human census.

Knowing of my former residence in New Zealand and my continuing fondness for that country, one of my students recently gave me an interesting (but undated) clipping from the Wall Street Journal telling about New Zealand's "flatulent sheep" (Brooks, n.d.). The New Zealand Institute of Nuclear Sciences had undertaken seriously to study, by radiocarbon dating, the age of various portions of methane in the air, and thus to ascertain how much is due to animals and how much emanates from fossil fuel sources, volcanoes, etc. A geophysicist with the Institute is quoted: "If you could hook up a sheep to the carburetor of your car you could run it for several kilometers a day." Somewhat indelicately he adds that to power the same vehicle from an equivalent human-generated fuel would require a whole football team "and a couple of kegs of beer."[4]

Quantifying the Problem

The image of methane-emitting livestock, together with the image of 2.25 billions of human equivalents populating the U.S. (some on two legs, some on four), brings us to a means of quantifying the statement about destruction of our survival base—in the case of the United States. Not only has the nation expanded across the continent since its birth. Not only has the number of citizens increased many-fold since the first U.S. census. In addition, the United States has "progressed" from being a mainly rural, agrarian society at the start to being an industrial behemoth today. What measure can express the magnitude and portent of that change?

Visualize a feed lot. Imagine within its confines not just a herd of fattening cattle, but a population of dinosaurs. Imagine the size of the lot growing as the decades pass, incorporating additional territory, but visualize at the same time both the number and the size of its dinosaur occupants increasing even more conspicuously. Then contemplate the enormity of the waste they would be accumulating underfoot!

The "issues of toxic waste disposal siting" would be vexing indeed if the "in which" environment lived in by these creatures was coterminous with the "into which" environment forced to accommodate the huge quantities of dinosaur excrement. That congruence of environmental uses does tend to be the case on a feed lot. And it is increasingly unavoidable on a postexuberant world with five billion variously equipped people aboard.

A city, as we said, has to have a "from which" and an "into which" hinterland, but as we convert our world into one huge city-like system we face problems arising from there being no hinterland for the planet as a whole.

Let us specify the size of our imagined dinosaurs (so they can serve as a measure of the load we Americans impose on our continental feed lot). Recent studies have produced a strong case for endothermy in dinosaurs—warm-bloodedness with its associated high metabolic rates. (Cold-bloodedness and low metabolism were originally attributed to dinosaurs when they were regarded as merely oversized relatives of present day lethargic reptile types—see, e.g., Desmond, 1977, especially pp. 80–85, 102–115, 171–172, 292.) If dinosaurs were endotherms we can apply to them a formula that exists for warm-blooded animals enabling us to calculate the body size that would be associated with a given rate of energy consumption, or vice versa (Kleiber, 1947). Dividing the daily expenditure of kilocalories by 70 and taking the 4/3 power of the quotient gives an estimate of the size of animal (measured in kilograms) expected to consume at that rate.[5]

Using total energy consumption as our measure of the human load a society imposes (via both the bodies and the technology of its citizens) upon its environment, it is then possible to depict this load in "dinosaur equivalents," as shown in Figure 6-1. During the 190 years from the first census to the latest, the U.S. "feed lot" expanded in area by a factor of four. The number of its occupants *per unit of area* increased during that time by a factor of nearly 13, but their average "size" (*per capita resource appetite and environmental impact*) *increased by a factor of 41.* Hence the contrast between the upper and lower parts of the diagram.

Although the correlation between per capita energy use and per capita generation of toxic wastes is presumably imperfect, it is surely positive and surely not negligible. The more energy at our disposal the more things we can do (Cottrell, 1955). The more things we undertake to do, the more we transform resources into wastes. When each of us has become effectively 41 times "as large" as our ancestors, and we are about

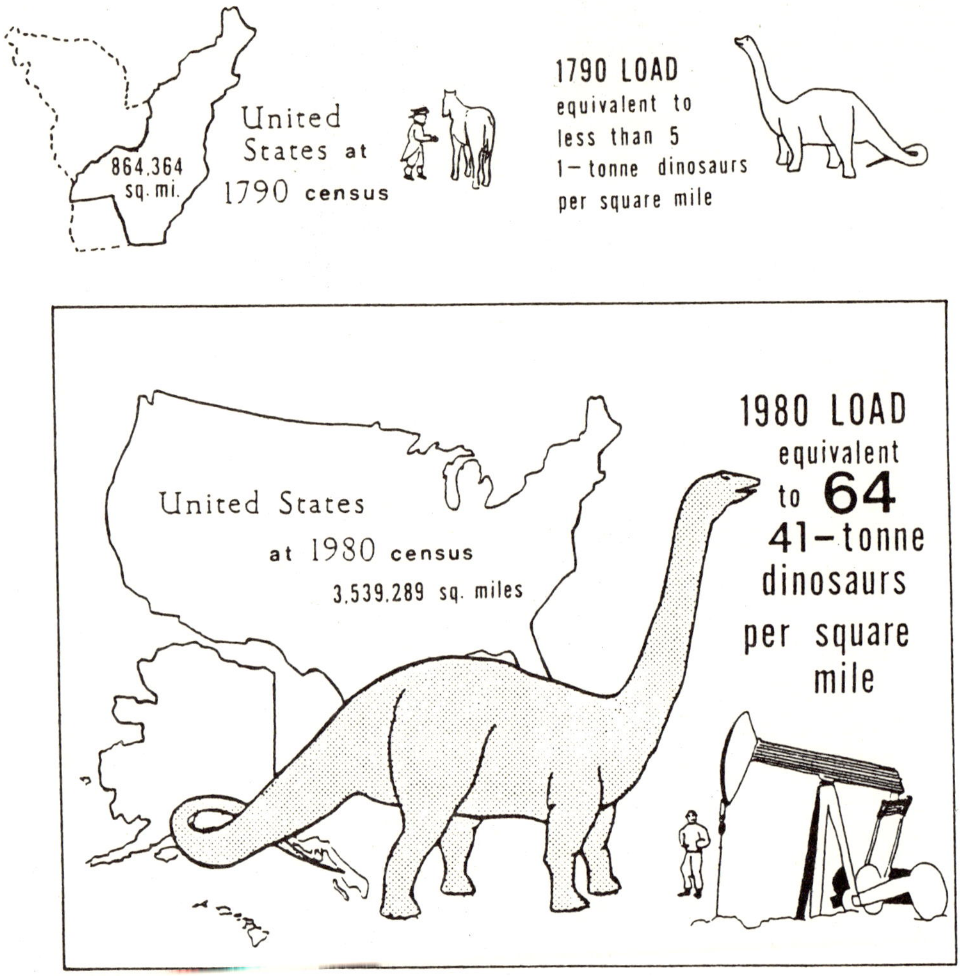

Figure 6-1. Two-way growth of metabolic load.

13 times as densely packed into our habitat as they were into theirs, it has ceased to be possible to keep the "into which" function of environment sufficiently segregated from the "in which" function. Hence my opening statement, here repeated, that the hazards in our children's lives will not be less than the ones we today keep trying to deny or downplay.

Disposal of hazardous waste has become a problem no longer soluble (in global terms) by site segregation.[6] Thus, in effect, we have imposed a new kind of "social contract" upon posterity; implicitly the price of being born into our world is "acceptance" by our descendants of a global "General Release of Liability." We are all living in a global Times Beach.

Mistakes can make our plight worse than it has to be, but there is no way we can escape the modern predicament short of reducing population and forgoing vaunted progress. This does not mean we should all throw up our hands in despair. It means we *must* understand what we can change and what we cannot. It means we all have an urgent obligation to make as few mistakes as is humanly possible.

ENDNOTES

1. In the more sophisticated science developed since the time I was in high school it is recognized that matter is not quite altogether indestructible. It can be, in special circumstances, transformed into energy. For our purposes here, however, this is an utterly negligible exception. It has been pointed out, for example (Gale and Hauser 1988), that to release the energy that destroyed Hiroshima an amount of matter equivalent to "one butterscotch lifesaver" was converted.
2. The importance of this distinction between three functions or uses was elucidated in a paper I read before the Natural Resources Research Group at annual meetings of the Rural Sociological Society (Catton and Dunlap 1986).
3. The increasingly prosthetic nature of modern man is perhaps epitomized in an illustration appearing with a newsmagazine article explaining how such tragic events as the shooting down of a civilian airliner by a U.S. Navy ship in the Persian Gulf could occur. Captioned "Combat in a black box: Computerized pilot's helmet," the picture showed a suited-up future pilot in a fighter plane cockpit. The huge helmet covering his head made him look like a fantasized extraterrestrial creature; the picture strongly suggests a "prosthetic head" or at least a prosthetic *extension* of the human head. According to the writer (Barry, 1988), "Instead of using their own eyes to find their targets, fighter jocks of the future will watch radar and infrared signals projected on the inside of the helmet visor. Like most such gadgets, the sensor-packed helmets are made necessary by the increasing speed and range of the opposition's weaponry."
4. To New Zealanders "football" means rugby, so the metaphor involves not eleven men but fifteen.
5. If these prehistoric species *were* cold-blooded, and hence less active, with less-than-mammalian rates of resource consumption and waste generation per kg of body mass, then the tonnage numbers given in Figure 6-1 would be gross *under*estimates of the "dinosaur equivalent size" of each American. However, the ratio between the 1980 and 1890 sizes would remain as stated. For criticism of the argument for dinosaurian endothermy, see Martin 1979.
6. If it does not seem this could be "an optimist's view," consider some matters I have left aside. Consider, for instance, the unlikelihood that our industrial civilization will curtail its generating of the principal environment-damaging waste whose emission is required by the material cultures needed to support our accustomed lifestyles—i.e., carbon dioxide. Meteorologist Bert Bolin, whose

encounter with the Scandinavian winter of 1942 aroused a lifelong interest in the effects of climate on human life, has said (Byrne, 1988): "To confront honestly the CO_2 problem means assessing our energy policies. How much coal, oil, and gas can we continue to burn in light of the mounting evidence of negative atmospheric effects?" We cannot, says Bolin, "hide in the bushes and count on nuclear energy to solve everything." And if CO_2 is a spectre haunting our future, consider the even more grim spectre I have chosen to disregard in this chapter, the plutonium modern technology has brought into the world in the past four decades, which, if scattered about, could be lethal to us all. As an ultimate instance of "cargoist" faith in yet-to-be-achieved technology, consider finally the prodigal accumulation of radioactive wastes by the nuclear power industry which went into production some forty years ago and the choice of a site for ultimate burial of this material is only now nearing resolution, with methods of safe storage still being studied.

REFERENCES

Barry, John.: "The Growing Risks of Fighting in the Dark." *Newsweek*, 112 (July 18):27, 1988.

Brooks, Geraldine.: "The 70 Million Sheep Roving New Zealand Create Quite a Stink." *Wall Street Journal*, no date.

Burch, William R., Jr.: *Daydreams and Nightmares: A Sociological Essay on the American Environment.* New York: Harper and Row, 1971.

Burford, Anne M., with John Greenya.: *Are You Tough Enough?* New York: McGraw-Hill, 1986.

Byrne, Gregory.: "The Winter of '42." *Science*, 240 (24 June): 1736, 1988.

Catton, William R., Jr.: *Overshoot: The Ecological Basis of Revolutionary Change.* Urbana: University of Illinois Press, 1980.

Catton, William R., Jr., and Riley E. Dunlap.: 1980. "A New Ecological Paradigm for Post-Exuberant Sociology." *American Behavioral Scientist*, 24: 15–47, 1980.

Catton, William R., Jr., and Riley E. Dunlap.: "Competing Functions of the Environment: Living Space, Supply Depot and Waste Repository. Paper read to the Natural Resources Research Group of the Rural Sociological Society, 29 August, Westin Hotel Utah, Salt Lake City, 1986.

CEK.: "Agriculture Poses Waste Problems." *Environmental Science and Technology*, 4: 1098–1100, 1970.

Council on Environmental Quality, and Department of State.: *The Global 2000 Report to the President.* New York: Penguin, 1980.

Cole, H. S. D., Christopher Freeman, Marie Jahoda, and K. L. R. Pavitt.: *Models of Doom: A Critique of the Limits to Growth.* New York: Universe, 1973.

Commoner, Barry.: *The Closing Circle: Nature, Man and Technology.* New York: Alfred A. Knopf, 1971.

Cotgrove, Stephen.: *Catastrophe or Cornucopia: The Environment, Politics and the Future.* Chichester, England: John Wiley, 1982.

Cottrell, Fred.: *Energy and Society.* New York: McGraw-Hill, 1955.
Day, Lincoln H., and Alice Taylor Day.: *Too Many Americans.* Boston: Houghton Mifflin, 1984.
Desmond, Adrian J.: *The Hot Blooded Dinosaurs: A Revolution in Palaeontology.* New York: Warner, 1977.
Dunlap, Riley E.: "Ecologist Versus Exemptionalist: The Ehrlich-Simon Debate." *Social Science Quarterly,* 64: 200–203, 1983.
Gale, Robert Peter, and Thomas Hauser.: *Final Warning: The Legacy of Chernobyl.* New York: Warner, 1988.
Garmon, Linda.: "Dioxin in Missouri: Troubled Times." *Science News,* 123 (January 22): 60–62, 1983.
Gough, Michael. *Dioxin, Agent Orange: The Facts.* New York: Plenum, 1986.
Hardin, Garrett.: *Exploring New Ethics for Survival: The Voyage of the Spaceship Beagle.* New York: Viking, 1968.
Kleiber, Max.: "Body Size and Metabolic Rate." *Physiological Reviews,* 27: 511–541, 1947.
Luten, Daniel B.: "The Limits-to-Growth Controversy." pp. 163–180 in Kenneth A. Hammond, George Macinko, and Wilma B. Fairchild (eds.), *Sourcebook on the Environment.* Chicago: University of Chicago Press, 1978.
Martin, Karen L.: "Thermoregulation in Dinosaurs and Pterosaurs." *The Biologist,* 61: 81–89, 1979.
Meadows, Donnella H., D. L. Meadows, J. Randers, and W. W. Behrens III.: *The Limits to Growth.* New York: Universe, 1972.
Merton, Robert K.: "The Unanticipated Consequences of Purposive Social Action." *American Sociological Review,* 1: 894–904, 1936.
Ozinga, James R.: *The Prodigal Human.* Jefferson, NC: McFarland, 1985.
Russell, Milton.: "The Process of Making Energy Choices." Part V. in Sam H. Schurr, Joel Darmstadter, William Ramsay, Harry Perry, and Milton Russell, *Energy in America's Future: The Choices Before Us.* Baltimore: Johns Hopkins University Press, 1979.
Simon, Julian L., and Herman Kahn.: *The Resourceful Earth: A Response to Global 2000.* Oxford: Basil Blackwell, 1984.

Chapter 7

COMMUNICATING RISK IN THE MASS MEDIA

ALLAN MAZUR

This discussion will focus on the communication of information about hazardous sites through the mass media, not in the sense of producing a well informed and calmly deliberative citizenry, but in the way it is usually done. Since 1978, journalists have become central actors in the public controversies over waste disposal that are now commonplace. No longer can refuse be quietly "dumped" on a passive community, for any such plans now make headlines, and some residents inevitably fight back, kicking and screaming to put the junk somewhere else. Most of us would react the same way if a hazardous dump were proposed for our neighborhood.

My thesis is that the *quantity* of news coverage about a waste site, rather than substantive details of its content, is the primary vehicle of communication about waste sites, and that most public impact follows directly from high media coverage. This position fits into the "agenda setting" tradition of research on mass media, which argues that news media, by giving heavy coverage to selected issues, raise these issues to public attention and concern (McCombs and Shaw, 1972). We go farther here, showing that high news coverage of a waste site not only raises public concern, but pushes it toward opposition to the site, raising fears in the adjacent community (Edelstein, 1988) and generally working against the siting of waste disposal facilities.

Sometimes this communication serves the public good, alerting the citizenry to real hazards and mobilizing it to seek solutions. At other times media coverage is detrimental to the public good, frightening people when no real hazard exists, increasing the costs of siting, or blocking useful disposal facilities. These, of course, are the two edges of the sword of NIMBY ("Not In My Backyard"), or of LULU ("Locally Unwanted Land Use"). While acronyms come and go, the dilemma remains: On the one hand, people must be able to protect their families and their communities from noxious and unfair intrusions; on the other hand, the society requires that some unwanted facilities be sited somewhere. The mass media may be helpful in resolving this dilemma or they may worsen it.

NIMBY

I first saw the slogan "not in my backyard" during 1969 when people were protesting the placement of antiballistic missiles in their neighborhoods. No doubt the phrase had been coined earlier, perhaps in the early protests against nuclear power plants or against the fluoridation of community water supplies. All these are good examples of local protests over environmental or technological issues that are at the same time national in scope, occurring nearly simultaneously at several places across the nation. Each local protest is linked to others in a more or less formal coalition, sharing newsletters and other literature, speaking the same rhetoric, and supporting nationally prominent spokespeople for the cause. The proponents of these various projects also form a united front so that the battle is joined not only at local sites but in the national arena as well. Most people watch these controversies through the lens of the mass media, having no direct role as the action moves among legislatures and executive offices of government and the courts.

While NIMBY (or LULU) controversies have existed for a long time, they have increased in frequency, achieving in 1987 the status of a topic heading (NIMBY Syndrome) in the *Readers' Guide to Periodical Literature.* There have been recent neighborhood protests against supermarkets, foster homes for infants, halfway houses for mental patients or substance abusers, and all kinds of other "undesirable" facilities (Hornblower, Sachs and Willworth, 1988). Waste disposal sites represent a prime target for these protests.

Several reasons for this increase exist. First, the need for these facilities has increased with our growing population, and it is no longer permissible to foist them on poor, politically important neighborhoods or rural communities. Also, there has been a marked increase in the number of citizen groups concerned about particular issues such as peace or arms, schools, pollution, abortion, or electoral reform. Encouraged by activist legislation such as the National Environmental Protection Act of 1969, and favorable administrative rulings, these special interest groups have more power than before to impede projects in the courts. Furthermore, such actions are now so well known in our culture that they provide a model for aggrieved citizens to follow should they be inclined to protest.

All of these factors have converged to make waste disposal sites a common target of community protest, especially since 1978 when the mass media first brought Love Canal to national attention, making its name a synonym for hazardous waste disaster. Prior to that time, few residents of this community, adjoining Niagara Falls in Upstate New York, knew it as "Love Canal." That name was popularized by the press, apparently for its ironic association with the Honeymoon Capital of America, and with the suggestion that residents had been screwed by the polluter, Hooker Chemical Company. It is worth assessing the role played by journalists at Love Canal to evaluate how risk communication works in practice, and how toxic waste reached the public's agenda of problems to be solved.

LOVE CANAL

A newspaper published near Love Canal, the *Niagara Gazette,* first began reporting possible hazards at the abandoned chemical dump site in 1976, when heavy rains caused buried chemical wastes to surface above ground. The story was initially ignored, but then taken up again in February 1978 by Michael Brown, a young *Gazette* reporter who had written several articles on another waste dump in the area. As a result of Brown's articles, the U.S. Environmental Protection Agency (EPA) and the New York State Department of Health began health and environmental studies. By May 1978, it was concluded that a serious threat to the health of persons living near the edge of the old dump existed. The *Columbia Journalism Review* later noted that Brown had "delineated a story of national significance. But small newspapers do not determine such matters" (Swan, 1979). It required *The New York Times* to bring the matter to national attention.

National Coverage

In late July 1978, Donald McNeil, Jr., a 24-year-old reporter, still in training, with an interest in environmental problems, arrived to investigate. Word of a story at Niagara Falls had already reached Sidney Schanberg, then metropolitan editor of *The New York Times*, through connections with members of the New York state legislature, which at the time was holding hearings on Love Canal. Schanberg was especially interested in the toxic chemical issue and, by his account, wanted someone young and energetic to cover the story. In McNeil's words, "Schanberg looked around for someone young and unmarried to send up to Buffalo, and it was me."

McNeil had never heard of Michael Brown or his articles before meeting him the first day at Love Canal. McNeil stayed at Brown's home that night, and the two men, of similar age and interests, became friends. Brown introduced McNeil to his sources among the homeowners, and they shared a common view of the events being reported.

By chance, McNeil arrived shortly before New York Health Commissioner Robert Whalen announced his department's findings and recommendations to the Governor. McNeil's first story, which appeared on the front page of *The New York Times* the morning of Whalen's press conference on 2 August, told the history of Love Canal and cited anecdotal accounts of numerous maladies suffered by residents, implying that these were caused by the chemicals. Only a few reporters, including Brown but not McNeil, attended Commissioner Whalen's press conference in Albany. The Commissioner's announcement of "a great and imminent peril" at Love Canal, and the recommendation that pregnant women and infants living at the edge of the dump move as soon as possible, coming right after the morning article in the *Times*, was national news. ABC and CBS television news reported on this event that same night. McNeil, who had been ready to return to New York, stayed on and produced stories every day, often on the front page, until 10 August when the *Times* stopped publication because of a strike. Love Canal had finally saturated the national news organs.

Love Canal's first appearance before a national audience lasted approximately one week, from McNeil's first story until the strike at *The New York Times.* By then, Governor Hugh Carey of New York and President Jimmy Carter had made emergency funds available for the evacuation and purchase of the 239 houses closest to the Canal. Although the purchase was announced as a direct response to the recently publicized "peril," it went far beyond the Health Commissioner's recommendations of a few days earlier. The surge of publicity throughout the country almost certainly contributed to this solution, especially as Love Canal immediately became an issue in the gubernatorial campaign for the primary election then in progress (Levine, 1982). With the purchase of the houses accomplished and *The New York Times* on strike, attention to Love Canal rapidly receded.

Battle for the Outer Rings

The mass media had helped the homeowners win their first battle, but there would be two battles, and two fairly distinct stories. The first concerned the houses in the inner ring, those closest to the Canal where clear signs of chemical seepage were found. This was the story reported by Brown in the *Niagara Gazette* and by McNeil for *The New York Times.* It was essentially settled a week after Commissioner Whalen's press conference by arrangement to purchase the houses.

By this time, residents of the outer rings, further from the Canal, were concerned that they too might be at risk. Led by Lois Gibbs, a minor character in the first battle, many of the remaining homeowners waged a new campaign, largely through the mass media, for their own houses to be purchased. Whereas in the first battle the homeowners saw the state's Department of Health as an ally in reaching a solution, in the second battle they perceived the Department of Health itself as the problem, a bureaucracy whose inertia and defensiveness became an impediment to the buyout of the remaining homeowners (Levine, 1982; Gibbs, 1982).

Various measures of both national and local news coverage show that attention dropped substantially after August 1978, when the problem of the inner ring was settled, but a lower level of attention remained during 1979, while remedial construction was under way at the Canal to confine the wastes (Fig. 7-1). Gibbs was effective in holding the attention of the press during this period; she became the major source of news among the homeowners and appeared on Phil Donahue's TV "talk show" and elsewhere, but she did not achieve the permanent relocation the homeowners sought.

The climax of the battle came on 17 May, 1980, on the front page of *The New York Times,* when Irvin Molotsky reported that 11 of 36 recently tested residents of Love Canal "may have suffered chromosome damage from toxic chemicals buried there." Based on an exploratory study that had been leaked to Molotsky by unnamed sources, the article further suggested that such damage is linked to cancer and genetic defects in offspring. The study soon became a focus of intense scientific criticism, particularly for its lack of a proper group, which left no way to determine whether the frequency of observed aberrations was abnormally high (Kolata, 1980).

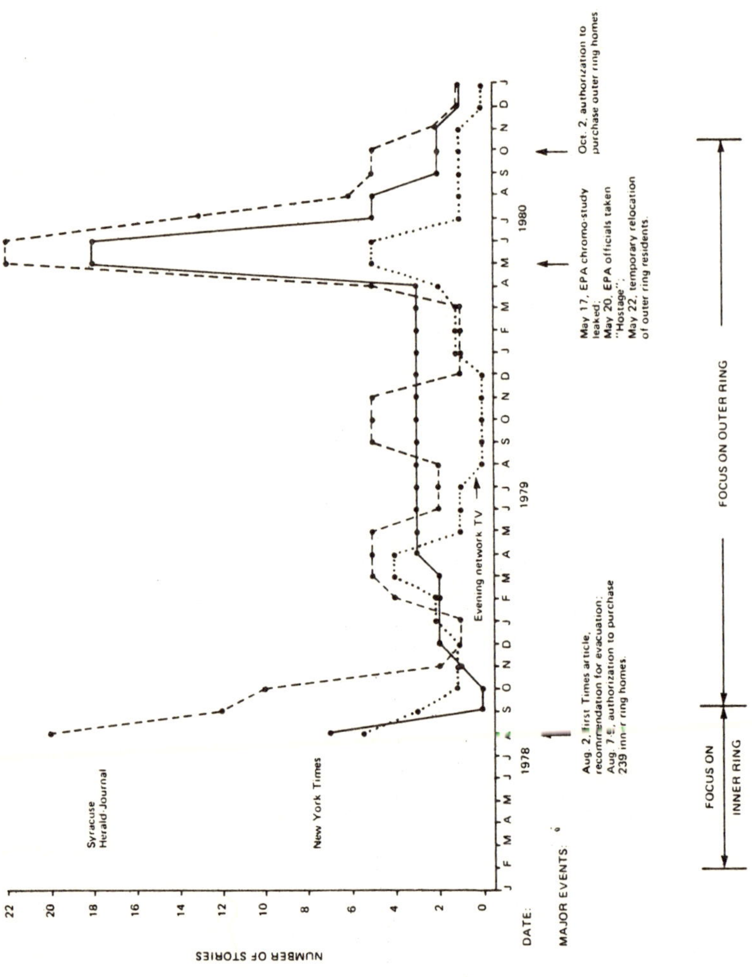

Figure 7-1. Number of stories on Love Canal in selected news media (smoothed by running medians).

A subsequent study with proper controls has since reported no excess of chromosome damage among Love Canal residents (Boffey, 1983). Molotsky had been given only fragments of the first study by his sources, and he more recently stated that, had he known there was no control group, he would have given less credence to the findings. Since the results of the initial study had been leaked to the Buffalo press, they would have appeared even if Molotsky had withheld his story, though without the authority of *The New York Times.* This dramatic story was widely disseminated and in my own city, Syracuse, NY, it was published under a front-page banner headline (Associated Press, 1980).

That Saturday afternoon the EPA held a press conference to discuss the study, and to announce that the findings of chromosome damage might make it necessary to relocate the 710 families still living at Love Canal. A decision on the move was to be made based on a review by geneticists of the chromosome study.

Homeowners at Love Canal, who had gone through two years of worry about the health of themselves and their families, were exceedingly distressed by the chromosome story, particularly those who had been tested but had not yet been informed of the results. As staff members of the EPA explained the tests, and as journalists reported the emerging story, frustration and anger grew in the community. On Monday, 19 May 1980, amidst a crowd of homeowners and reporters, Lois Gibbs took two officials of the EPA "hostage," telephoning this information to the White House (Gibbs, 1982). At this time, American diplomats were being held hostage in Iran.

The interest of press and public in toxic chemicals, which had been growing since 1976, stimulating attention to Love Canal and in turn being stimulated by it, was in 1980 higher than it had ever been (Mazur, 1984). With this heightened receptivity, the events of May 17 to 19 received more journalistic attention than was given to Love Canal at any previous time. By Wednesday, 21 May, President Carter declared a state of emergency, although his advisers apparently regarded the chromosome study as invalid, permitting the federal government and the state of New York to relocate temporarily the families from the outer ring. Once the residents had been moved from the area, the purchase of houses, approved that October, became inevitable. With the help of the mass media, the homeowners had won the second battle.

The Public Good

At Love Canal, media coverage worked to the advantage of those protesting corporate and government actions. The homeowners' concerns quickly moved from the local to the national level. Newspapers and television, which heretofore had given little attention to waste problems, now made them a major issue of environmental news, placing toxic waste on the nation's agenda. Congress, spurred by the Love Canal publicity, quickly created the Superfund, administered by the EPA to finance the cleanup of toxic waste sites around the country.

Most observers apparently believe that the heavy media coverage of Love Canal served the public good, both in relieving the plight of the endangered homeowners and in bringing the important issue of hazardous waste disposal to national attention. However, from another perspective, we may see the Love Canal controversy as a battle for scarce public resources, where each side's claims have some face validity (Hammer et al., 1980; Zuesse, 1981). While no one denies that toxic chemicals were seeping into homes adjoining the Canal, it is doubtful that they caused the degree of harm claimed by the homeowners. Most early claims of chemically induced illness were either anecdotal, such as those cited in *The New York Times;* or they were based on informal surveys by the homeowners themselves (Levine, 1982; Gibbs, 1982). A New York Department of Health report, released in September 1978, was the most credible of the early claims of damage, suggesting excess miscarriages, birth defects, and babies with low birth weight, all were attributable to chemicals leaking into homes near the Canal (*Love Canal: Public Health Time Bomb,* 1978).

A committee of experts, appointed by Governor Carey to review most of those early statements, concluded that they were of little scientific value and did not establish relationships between health and the hazardous wastes at Love Canal (Thomas et al., 1980; see Levine, 1982, for a critique of this report). By 1981 the New York Department of Health had moderated its own early claims of fertility problems, regarding the evidence of excess birth defects as ambiguous. The Department of Health reaffirmed that birth histories of Canal residents showed a "slight excess" in miscarriages but pointed out that this excess had occurred in the decades prior to 1970 and was not apparent afterward (*Love Canal,* 1981). The incidence of babies with low birth weight was abnormally high during the 1950s, the period of active dumping, but babies born after

1960 were at no greater risk of low birth weight than were those born in upstate New York (Vianna and Polan, 1984). Furthermore, data from the New York Cancer Registry show no evidence for higher cancer rates associated with residence near the Love Canal site in comparison with the entire state outside of New York City (Janerich et al., 1981), nor do ex-residents have more chromosome abnormalities than properly designed control groups (Boffey, 1983). It is possible that some illness has escaped detection, or that cancers with long latency periods have not yet appeared. Nonetheless, the most credible evidence to date shows little damage to health during the 1970s from leaking wastes.

No one doubts that toxic chemicals seeped into homes bordering the Canal, but a combination of journalistic coverage and the alarmed early reaction of the New York Department of Health seems to have exaggerated the level of hazard to the community as a whole. Was the public good really served by diverting so much public money in 1978 and again in 1981 to purchase nearly 950 homes from families whose fears by then we can surely understand, even if most of them were not at excessive risk from chemical waste? The EPA itself has taken the position that the amount of money devoted to the Superfund is far out of proportion to the actual risk from toxic waste sites, and that other hazards, far worse, receive much less funding for alleviation (U.S. EPA, 1987).

PROTESTORS AND MEDIA COVERAGE

I propose that media coverage helped the Love Canal homeowners, not because their case won on its merits, but because media coverage *almost always* works to the advantage of the protestors in environmental cases, whether the merits are on their side or not.

Protesters and journalists have a symbiotic relationship, each serving the interests of the other. Consider the plight of a reporter assigned a story about PCB leakage from a waste site. How much of interest is there to write or photograph? PCB-laced dirt does not make much of a television picture. The obvious way to improve this story, for journalistic purposes, is to add human interest by showing suffering or controversy. If there is a protest meeting, especially a vitriolic one, that makes good news copy and pictures. Often the reporter will seek out a family living near the site, so that television can show a mother and her children in fear of contamination. Especially photogenic or articulate mothers are sought out repeatedly by journalists, as was the case with Lois Gibbs at

Love Canal, turning them into local celebrities empowered to lead other protestors.

Reciprocally, protestors thrive on media coverage. It is a heady experience to see oneself on television or to be quoted in the newspaper; such notoriety encourages further activism. Most protest groups depend heavily on volunteer workers. Like all of us, these volunteers have other demands on their time. When confronted with a range of worthwhile but competing activities, they usually choose the one where the action is, the one that is receiving most publicity and is most caught up in public controversy. Every special interest organization that depends on volunteers' work and monetary contributions knows that donors are easiest to find when the organization and its cause are being publicized; when the publicity ends, donations dry up.

Protest activity and media coverage are tightly linked, going up and down together. Protest is newsworthy, especially when it is outlandish like Lois Gibbs' "hostage taking." At the same time, publicity spurs the protest and keeps up its momentum.

The strategic lesson here for protestors is to strive for as much media coverage and confrontation as you can get. Those on the opposite side should downplay coverage and controversy, hoping that the protest will lose strength and fade away.

MEDIA EFFECTS ON PUBLIC OPINION

The number of people really active in any protest is only a small portion of the community. Most people are fairly passive, only registering their views in the voting booth or on an opinion poll. Still, by shifting its opinion for or against a proposed waste site, this wider public can alter the climate within which activists operate, making it friendlier or more hostile to them.

In the long controversies over nuclear power plants and fluoridation, public opinion for or against these technologies has fluctuated with the amount of media coverage they have received. When media coverage increases, then public opinion tends to become more opposed; as media coverage declines, then public opposition drops off, too. This relationship between *quantity* of coverage and intensity of opposition holds even when editors are generally favorable toward the technology, as in the case of fluoridation (Mazur, 1981).

This is a puzzling result since one would expect public opinion to turn negative or positive depending on whether the media coverage was generally unfavorable or favorable in content. Perhaps if news coverage were wholly favorable then increases in coverage would indeed be followed by increasing positive opinions. However, news coverage of controversial technologies is never wholly positive because of the norm in American journalism to provide balance in stories about controversial topics. The reporter must include both sides of the debate, interweaving comments for and against the technology, finally leaving readers or viewers to draw their own conclusions. Thus, as a practical matter, increased coverage of a controversial topic always includes an increase in derogatory as well as favorable comment.

If coverage is balanced, why should there be a net shift toward opposition? Would not each positive comment simply cancel a negative one? This may have been the reasoning of the Media Task Force of the President's Commission on the Accident at Three Mile Island (1979; Rubin, 1981), which sought to evaluate charges that news reports of that nuclear reactor accident were overwhelmingly sensationalist. The Task Force counted the number of alarming or reassuring statements in a sample of news reports during the week following the beginning of the accident. Since there were more reassuring than alarming statements (56 to 39 percent, with 5 percent neutral), the Task Force concluded that the press did not portray the accident in an alarming manner. The problem with this reasoning is easily shown with a hypothetical report like "the reactor may explode, but it probably will not," which remains distressful despite having one alarming statement balanced by a reassuring one.

Once a hazard is made salient, even in a qualified way, people may become very worried. In an experiment by Mueller (1968), subjects who were exposed to both positive and negative arguments about fluoridation were more likely to oppose it than people who had not seen any of the arguments at all. Fluoridation is more likely to be defeated in a referendum when both sides of the argument have been heatedly debated than when the campaign has been relatively quiet (Crain et al., 1969). During a local controversy over a proposed nuclear waste storage facility, residents who had heard about the controversy were more likely to oppose the facility, and to consider it unsafe, than residents who had not heard about the controversy (Mazur and Conant, 1978).

The negative attitudinal change that a balanced article may produce in the mind of a reader is illustrated by college students who were shown a high-quality news article about water fluoridation (Haas, 1980), a public health measure that most students initially favored while claiming to know little about it. This long article was balanced, beginning with a section on the benefits and safety of fluoridation (#1-pro), then reporting charges by some critics that fluoridation causes cancer (#2-anti), then telling of profluoridationists' refutations of these critics (#3-pro), then reporting charges by "more credible" critics that fluoridation is a hazard (#4-anti), and concluding with a statement from profluoridationists that the water treatment is safe (#5-pro). The article's sections were projected in turn on a screen in front of about 100 students. As the students read each section, they recorded their attitude toward fluoridation as favorable, opposed, or neutral. They had also recorded their initial views, before seeing the article, and they did so once again six weeks after the screening.

Figure 7-2 shows the evolution of student views, from their predominantly profluoridation stance before seeing the article (9 percent opposed, 47 percent in favor), through vacillation as they read the various pro- and antifluoridation sections, until at the end of the article nearly as many opposed (42 percent) as favored (44 percent) fluoridation. After six weeks, profluoridation views were reasserted, but the ratio of opposition votes to favorable votes was still higher than it had been initially. The shift toward opposition was especially high among students who had been initially neutral toward fluoridation. If a single balanced article can produce this effect, a sustained flow of balanced news coverage should ensure that the shift toward opposition persists, at least as long as the coverage persists.

Simple Images

Too often analyses of mass media communication assume that the reader or viewer gives undivided attention to the message, absorbing it in its entirety and extracting from it what the reporter put into it. In fact, viewers often watch television news with half an eye over the dinner table, and hardly anyone carefully reads the whole newspaper, instead skimming the pages for articles of special interest while barely glancing at many others. Therefore we must distinguish between the substantive content of a story, which should be intelligible to a careful reader or viewer, and the simple image that most casual consumers actually absorb from the story, for these may differ.

While the intent of a carefully balanced story about a controversial landfill may be to fully and fairly inform the reader of all sides of the

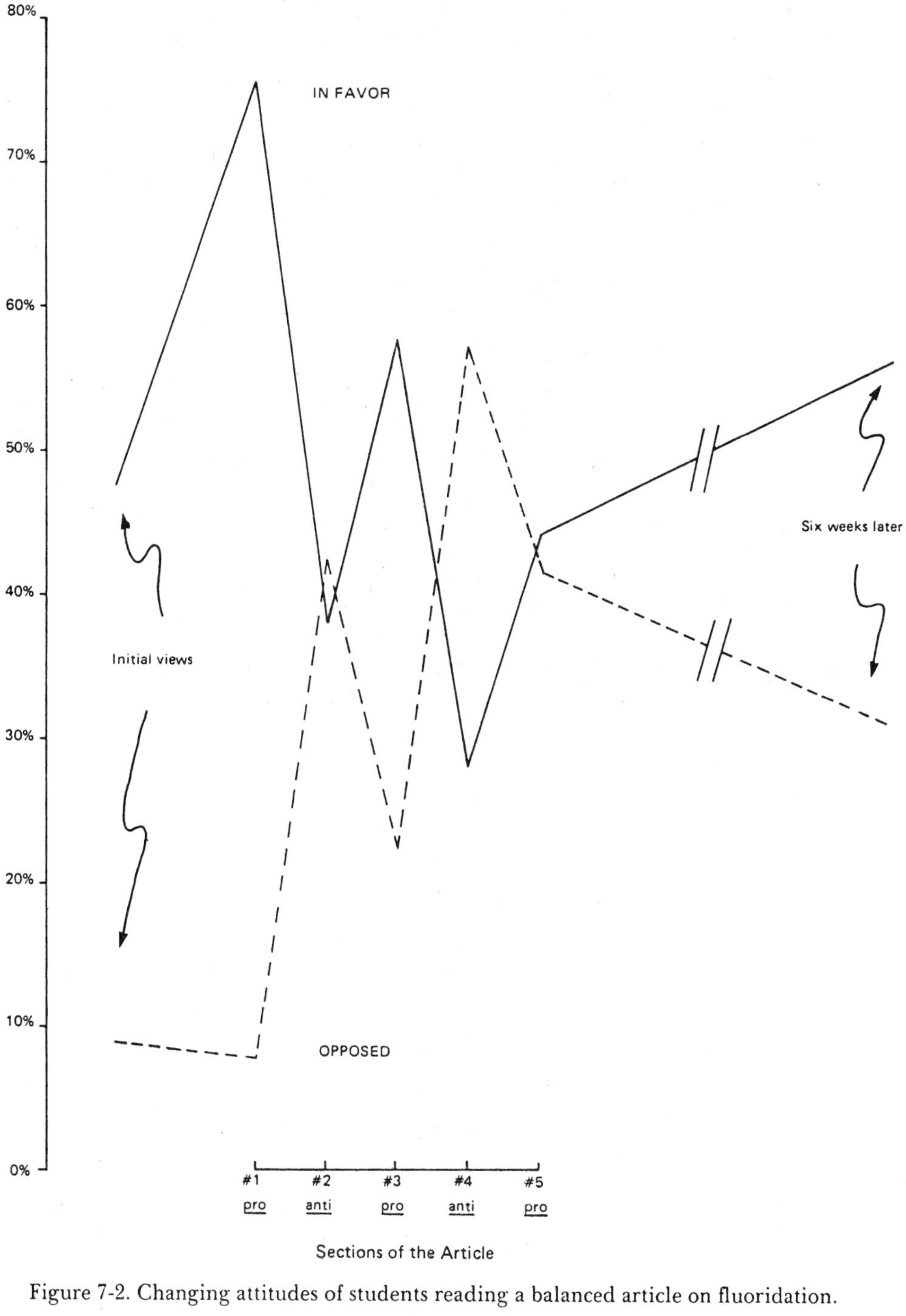

Figure 7-2. Changing attitudes of students reading a balanced article on fluoridation.

debate so that he may approach it with an open mind, the simple image of the landfill that is actually conveyed to the average (casual) reader may be quite different. First, the balanced presentation of opposing sides suggests that each position is equally credible, even when one side represents the mainstream of technical expertise while the other is the position of a few dissidents who may or may not have a competent case. Furthermore, the mere suggestion of negative outcomes, even if balanced by positive ones, evokes more caution than if no outcomes are mentioned at all (Tversky and Kahneman, 1974). If doubt is raised about the safety of the technology, many in the public prefer to err on the side of safety, as if saying, "When in doubt, reject the technology—better safe than sorry" (Sapolsky, 1968).

The casual newspaper reader may get his impression of a story from its headline, an accompanying picture and its caption, and perhaps the first paragraph or two of text. Unlike a careful perusal, this quick but typical skimming exaggerates the importance of the simple image that is conveyed by the overall impression of the article on the page, or by the visual forms on the TV screen.

As an example, the Sunday newspaper in my own city of Syracuse, New York, devoted more than one-third of its front page, plus two and one-quarter inner pages, to New York State's search for a disposal site for low-level radioactive waste (*Herald-American;* January 22, 1989). The state had recently announced a list of ten potential sites, and the newspaper focused on those within its circulation area, especially discussing incipient public opposition at each site, nearby producers of radioactive waste, and general information about the waste.

Surely not very many people outside of the affected areas read the nearly three pages of waste-related stories. On the other hand, virtually everyone looking at the newspaper must have scanned the one-third of the front page with the red label "Special Report" over the bold headline, "NUCLEAR JUNK: Debate Over Dump Site." The dominant feature on the page is a large picture of a man in a yellow radiation-protection suit with a plastic hood covering his head. Two subheads read "Radioactive waste will be stored in state," and "Residents of nuke-waste sites unite to fight" (Fig. 7-3). Readers who turned to the inner pages found among the text three pictures of concerned citizens, another picture of a radiation physicist in white coat measuring exposure from a nuclear waste container, a diagram explaining the types and quantities of radioactive waste, and a map of the state showing the ten potential sites.

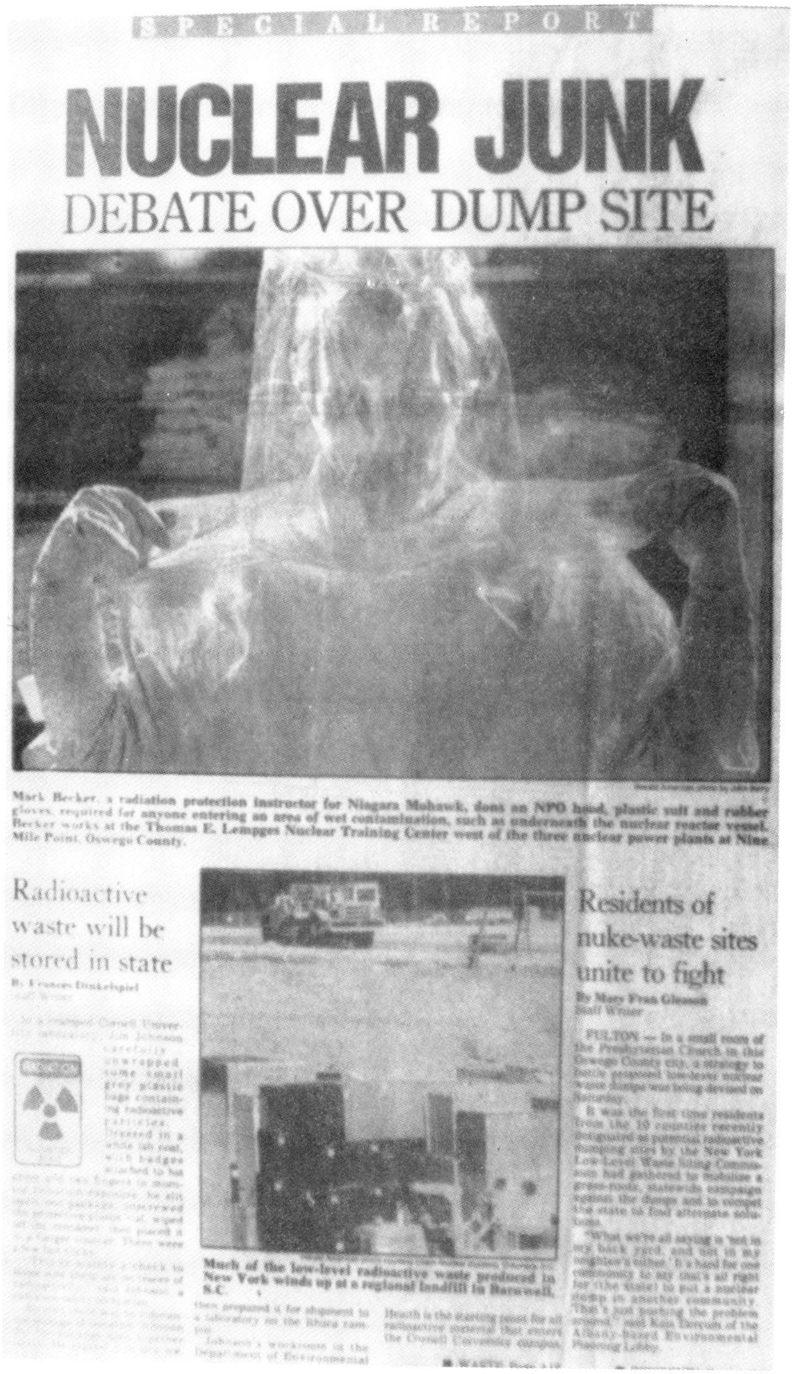

SPECIAL REPORT

NUCLEAR JUNK

DEBATE OVER DUMP SITE

Mark Becker, a radiation protection instructor for Niagara Mohawk, dons an NPO hood, plastic suit and rubber gloves, required for anyone entering an area of wet contamination, such as underneath the nuclear reactor vessel. Becker works at the Thomas E. Lempges Nuclear Training Center west of the three nuclear power plants at Nine Mile Point, Oswego County.

Radioactive waste will be stored in state

Residents of nuke-waste sites unite to fight

Much of the low-level radioactive waste produced in New York winds up at a regional landfill in Barnwell, S.C.

Figure 7-3.

The newspaper's managing editor, Timothy Atseff, explains the particular placement and packaging of the stories as intending to engage readers in important information on an emerging public issue. Since the newspaper reaches about 85 percent of country residents each week, there is little concern about selling more papers but rather with attracting the attention of the present readership, which is better done with a focus on controversy, a bold headline, and a dramatic picture of the yellow-suited man, than with a pictureless clinical presentation of the same information, such as one might find in the *Wall Street Journal.* Atseff perceives his readers as becoming better informed by the stories without experiencing any affective tilt one way or the other.

Taken as a whole, the collection of stories is balanced, explaining on the one hand the medical and industrial uses of radioactive material, and the need for waste disposal, while on the other hand describing the hazards of radiation and telling of citizen opposition to a waste facility. Yet it seems inescapable that the overall impression created by this collection of articles is a negative one.

Worse yet is the simple image that most readers obtain from a rapid scan of the photos and headlines.[1]

MASS MEDIA COMMUNICATION AND THE PUBLIC GOOD

Stories about the low-level radioactive waste sites followed quickly on local television stations, reflecting the importance given to it by the newspaper, which typically sets the agenda for the Syracuse news community. Public meetings held at the various potential sites during the following week were well attended by protesters and members of the wider public, sometimes numbering in the thousands, many of them frightened (Tonello, 1989; Associated Press, 1989), no doubt in part the results of heavy media coverage. Few observers give New York State much chance of opening a low-level radioactive waste facility by 1993, the date Congress has set for each state to have its own disposal site rather than shipping low-level waste out of state.

---

[1]I am indebted to Professor W. Henry Lambright of Syracuse University for eliciting from students in his graduate seminar on science and public policy their written impressions from a rapid exposure to the front page. Eight of the 15 students made explicitly negative characterizations, their most striking images of the yellow-suited man being "Andromeda Strain," "ghoulish," "death hermetically sealed," "scare people about nuclear waste (radioactivity) with the jaundiced, stretched skin." Impressions by the other seven students were neutral in connotation; no one reported a positive image.

If we assume that it serves the public good for the state to provide a low-level radioactive waste facility, or a facility for any kind of hazardous waste disposal, then we must doubt that the mass media are serving that goal by producing a well informed and calmly deliberative citizenry, for that hardly occurs at all. The detailed substantive content of media coverage is not as much at issue as the sheer quantity of coverage, and the negative simple images that it produces in the public mind. Professor Peter Sandman of Rutgers University points out that journalistic coverage often has its most potent effect merely by generating outrage over who is responsible for a hazard, or who has lied about it. These simple images fuel public opposition and widespread fear independent of the merits of the issue. Surely there are times when fear and protest are proper responses, but today they have become inevitable responses, and that can hardly be proper all the time.

One cannot advocate that journalists stop covering these matters, which are clearly newsworthy issues of public concern. However, we can ask that the quantity of coverage becomes more routine and consistent rather than the current practice of turning it up or down in erratic fits and starts so that a burst of coverage is inevitably taken by the public as a signal of danger. We can also ask that more attention be given to the simple images that are conveyed by headlines and visual images, which may overwhelm and distort the substantive content of the communication. These are matters for the journalists themselves to take in hand, for with their First Amendment sensitivities, Americans are loathe to constrain the news media. At the same time we ought not to ignore the power that media institutions have to promote or retard the public good.

REFERENCES

Associated Press: Chromosome Damage Possible. *Syracuse Herald-Journal,* May 17: 1, 1980.

Associated Press: Upstaters scared over possibility they'll live near radioactive waste. *Syracuse Herald-American,* February 12: C2, 1989.

Boffey, Philip: New Love Canal study finds no links to genetic diseases. *New York Times,* May 18: A1, 1983.

Crain, R., Katz, E., and Rosenthal, D.: *The Politics of Community Conflict.* Indianapolis, Bobbs-Merrill, 1969.

Edelstein, Michael: *Contaminated Communities.* Boulder CO, Westview Press, 1988.

Gibbs, Lois: *Love Canal: My Story.* Albany, State University of New York Press, 1982.

Haas, M.: Fluoride debate rages: Poison or dental panacea? *The Syracuse Post-Standard,* December 29: A-10, 11, 1980.

Hammar, Armand, et al.: *The Other Side: Love Canal.* Los Angeles, Occidental Petroleum Corp., 1980.

Hornblower, Margot, Sachs, Andrea, and Willwerth, James: Not in my backyard, you don't. *Time* (June 27): 44–45, 1988.

Janerich, Dwight, et al.: Cancer incidence in the Love Canal area. *Science, 212* (19 June): 1404–07, 1981.

Kolata, Gina: Love Canal: False alarm caused by botched study. *Science 208* (June 13): 1239–42, 1980.

Levine, Adeline: *Love Canal: Science, Politics, and People.* Lexington, MA, Lexington Books, 1982.

Love Canal. Albany, New York State Department of Health, 1983.

Love Canal: Public Health Time Bomb. Albany, New York State Department of Health, 1978.

Mazur, Allan: The journalists and technology: Reporting about Love Canal and Three Mile Island. *Minerva, 22* (spring): 45–66, 1984.

Mazur, Allan: *The Dynamics of Technical Controversy.* Washington DC, Communications Press, 1981.

Mazur, Allan and Conant, Beverlie: Opposition to a local nuclear waste repository. *Social Studies of Science, 8:* 235–43, 1978.

Mazur, Allan: Controversial technologies in the mass media. Pp. 140–58 in M. Kraft and N. Vig. eds., *Technology and Politics.* Durham, NC, Duke University Press, 1988.

Maxwell McCombs, and Shaw, Donald Shaw: The agenda-setting function of the mass media. *Public Opinion Quarterly, 36:* 176–187, 1972.

Molotsky, Irvin: Damage to chromosomes found in Love Canal tests. *New York Times,* May 17: 1, 1980.

Mueller, John: Fluoridation attitude change. *American Journal of Public Health, 58:* 1876, 1968.

President's Commission on the Accident at Three Mile Island: *Report of the Public's Right to Information Task Force.* Washington, DC, U.S. Government Printing Office, 1979.

Rubin, David: What the President's Commission learned about the media. Pp. 95–106 in Thomas Moss and David Sills, eds., *The Three Mile Island Nuclear Accident,* New York, New York Academic of Sciences, 1981.

Sapolsky, Harvey: Science, voters, and the fluoridation controversy. *Science, 162:* 427, 1968.

Swan, John: Uncovering Love Canal. *Columbia Journalism Review, 17:* 50, 1979.

Thomas, Lewis, et al.: *Report of the Governor's Panel to Review Scientific Studies and the Development of Public Policy on Problems Resulting from Hazardous Wastes.* New York, Memorial Sloan-Kettering Cancer Center, 1980.

Tversky, A., and Kahneman, D.: Judgment under uncertainty: Heuristics and biases. *Science, 185:* 1124–31, 1974.

Tonello, John: Waste-site panelists bemoan role as 'flak-catchers'. *Syracuse Herald-American,* January 29, 1989.

U.S. Environmental Protection Agency: *Unfinished Business: A Comparative Assess-*

ment of Environmental Problems. Washington, DC, Office of Policy, Planning and Evaluation, February 1987.

Vianna, Nicholas and A. Polan: Incidence of low birth weight among Love Canal residents. *Science, 226* (December 7): 1217–1219, 1984.

Zuesse, Eric: Love Canal: The truth seeps out. *Reason* (February): 16–33, 1981.

Chapter 8

COMMUNITY PREPAREDNESS FOR HAZARDOUS WASTE EMERGENCIES[1]

CHARLES E. FAUPEL AND CONNER BAILEY

INTRODUCTION

> Someone once suggested that Noah, with his ark, was the first disaster planner. He anticipated a threat, having a somewhat unusual and personalized warning system. Certain consequences seemed probable.... He projected his manpower needs and had the capability to mobilize the necessary personnel. When the threat was realized, he rode out the storm in reasonable safety and, in not too many days, was ready to start on the recovery stage...
>
> (Dynes, Quarantelli and Kreps, 1981:1)

Contemporary emergency planners continue to confront many of the same planning problems and issues as Noah, though not always as successfully. Emergency management has become a profoundly more complex enterprise in subsequent millenia, owing largely to the growing complexity of society itself. Contemporary emergency management entails the coordination of a variety of emergency relevant organizations. Moreover, not only must contemporary emergency planners anticipate a multiplicity of potential natural hazards, but, unlike Noah, their job is further complicated by the threat of *technologically*-based emergencies as well. Technological progress has brought with it vulnerability to fires, explosions, radioactive contamination and catastrophic accidents (Perrow 1984).

Modern technology has produced one further side effect which has direct relevance to emergency planning—waste, and much of it of a hazardous nature. Until about a decade ago, disposal of such waste was handled in an ad hoc and piecemeal fashion as industrial plants and

[1]Field research and preparation of this chapter was supported by the Alabama Agricultural Experiment Station at Auburn University as part of a project examining public involvement in hazardous waste management.

other producers of hazardous waste simply dumped it in the most economically efficient way possible, often on their own premises.

It was not until 1965, with the Solid Waste Disposal Act, that Congress officially recognized that waste materials posed a threat to the environment. While this was an important piece of legislation symbolically, it was limited to providing research funds and technical assistance to state, county, and city planners. The Resource Recovery Act of 1970 expanded the substantive focus of the 1965 legislation, promoting the use of sanitary landfills, conservation measures, and recycling technologies. The threat of hazardous waste as a specific focal concern was overlooked until 1976 with the passage of the Resource Conservation and Recovery Act (RCRA). This piece of legislation, enforced by the Environmental Protection Agency (EPA), set minimal standards for the treatment, storage, and disposal of hazardous waste products. These RCRA standards, which were made even more stringent in 1984, regulate hazardous wastes produced by approximately 1,000 small quantity generators (i.e., those who generate 100–1,000 kilograms of hazardous waste per month) as well as approximately 3,000 large quantity generators (EPA, 1987). This latter group, which produce over 1,000 kilograms of hazardous wastes per month, accounts for 99 percent of the roughly 250 million tons of hazardous wastes produced and managed under RCRA (EPA, 1987). The majority of all hazardous wastes are handled on-site by large-scale generators. The operation of off-site commercial facilities, however, has become a major growth industry, particularly for the treatment and disposal of the most highly toxic waste streams.

The concentration of large amounts of hazardous materials in central locations poses two kinds of potential environmental threats. There are, first, *chronic threats* to human health and the environment. Groundwater and aquifer contamination, occasional atmospheric releases, and toxic exposure of workers handling these materials all constitute a sufficient threat to require on-going monitoring by federal and state regulatory agencies. In addition, there are *acute threats* which, while less frequent in nature, pose potentially catastrophic risks. Explosions may occur, for example, if two otherwise nonvolatile substances accidentally come into contact with one another. Such an incident not only endangers the life and safety of those in the immediate vicinity, but also potentiates the release of large amounts of toxic fumes into the atmosphere. The increased volume of traffic transporting hazardous materials on highways and waterways raises the specter of major accidents and subsequent release of large volumes of hazardous and toxic substances. Similarly, major on-site accidents which could endanger the health and lives of hundreds of workers and residents in the surrounding area, however unlikely, are not inconceivable.

Scenarios such as these must be anticipated and require careful contingency planning. Fortunately, decades of research have revealed a number of principles which are characteristic of successful emergency planning (Dynes et al., 1981). While these principles have evolved from studies of natural disasters, they are sufficiently generic that they may be applied to other types of emergencies, including those associated with hazardous chemicals (Faupel et al., 1987; Quarantelli, 1984). Beyond these general principles, however, there are a number of contingencies associated with hazardous wastes which introduce a unique set of social constraints to effective community preparedness for an emergency involving such materials.

This paper seeks to address these problemareas with some exploratory observations of emergency preparedness in Sumter County, Alabama, site of the nation's largest hazardous waste landfill. Sumter County is located in western Alabama's "Black Belt" region on the Mississippi border (U.S. Bureau of the Census, 1983). The county is largely rural with a population of 16,908, over 69 percent of which is black. Its largest concentration of population is located in the cities of York (pop. 3,392) and Livingston (the county seat, pop. 3,187), both located in the southern portion of the county. The waste disposal facility is located in the sparsely populated northern sector of the county, between the small communities of Emelle and Geiger. The facility was established in 1976 by Resource Industries, Inc., a small company owned and controlled by a group of regional investors. Chemical Waste Management, Inc. (popularly known as "Chem Waste") bought out Resource Industries' interests in 1978 and has operated the site since. Chem Waste is a subsidiary of the multinational Waste Management, Inc., which has its headquarters in Oak Brook, Illinois. Chem Waste is the world's largest company in the field of hazardous waste management.

DATA AND METHODS

This study is part of a larger research endeavor which seeks to assess the impact of the hazardous waste disposal industry on local community life. In the context of this broad research goal, the present analysis of local emergency preparedness is understood to be one measure of the extent to which potential negative impacts are mitigated.

Data comprising this analysis are drawn from semi-structured interviews with individuals in three distinct samples. The first "emergency official" sample consists of representatives from each of the most relevant emergency organizations in the county. A total of nine respondents representing ten organizations were interviewed.[2] These interviews, which lasted one to two hours, specifically addressed relevant emergency planning and preparedness issues in the county. Such issues include the nature and extent of interorganizational coordination; interorganizational conflicts and other problems encountered in coordination with related emergency organizations; perceived strengths and weaknesses in the planning process; perceived organizational domains; and perceived impediments to carrying out these domains. Finally, a "worst case scenario" of a chemical emergency was presented, and respondents were asked to indicate how they would respond to such a situation.

The second "knowledgeable public" sample consists of thirteen purposely selected individuals who were knowledgeable regarding local political, economic and community life. The sample was carefully chosen so as to represent differing population groups within Sumter County, including representatives of both black and white residents, people from both the southern and northern portions of the county, and local residents favoring as well as those opposed to the presence of the Chem Waste facility. Thirteen individuals were interviewed for two to three hours each. The interviews were broadly focused on the effect of the hazardous waste facility on political, economic and community life in the county. The third sample consisted of ten "community leaders" identified by the "knowledgeable" panel as being the key leaders within Sumter County. These community leaders also were interviewed for two to three hours each on the political, economic, and social impact of Chem Waste's operations on Sumter County.

[2]Organizations interviewed include: Alabama Highway Patrol, municipal police departments (2), fire departments (2), ambulance services (2), hospitals (2), the local Emergency Management Agency, and Chemical Waste Management. Attempts were made to interview representatives from the county sheriff's department and the all volunteer county rescue squad. Due to schedule conflicts, however, these interviews could not be conducted.

PROBLEMS IN EMERGENCY PREPAREDNESS FOR HAZARDOUS WASTES

In general, emergency planning principles developed for natural disasters can be applied to chemical emergencies. The need to plan for interorganizational coordination, to anticipate likely public response, and to periodically update and test formalized plans, is as important for response to chemical emergencies as it is for hurricanes, tornadoes, floods and other natural agents of destruction (Faupel et al., 1987; Quarantelli, 1984). However, the social context of planning for chemical emergencies is quite distinct from that of natural disasters.

Politicization of the Hazardous Waste Industry

It has long been observed that technologically-created disasters pose different dynamics than natural disasters. The locus of natural disasters is outside the affected social system. Consequently, communities struck by natural disasters perceive a common threat resulting in heightened community solidarity (Faupel, 1985). Such is not the case with technologically-caused disasters. In these instances the cause of the disaster is clearly within the social system itself and a process of assigning blame inevitably ensues (Bucher, 1957; Drabek, 1968; Drabek and Quarantelli, 1967; Veltford and Lee, 1943). Rather than producing heightened community solidarity, technological disasters usually promote conflict and exacerbate social and political cleavages in the impacted community.

Moreover, some technological hazards pose such a chronic threat that they engender on-going community conflict. Recent research on the nuclear power industry in the wake of Three Mile Island, for example, reveals the intensity of public reaction against such potentially devastating technology (Walsh, 1984). Similarly, the chemical industry (Nigg and Cuthbertson, 1982), and particularly the hazardous waste industry (Levine, 1982; Neal, 1984), provoke strong emotional responses from the affected community.

Sumter County is not immune from these characteristic conflicts. Almost from its inception the facility at Emelle was met with grass roots resistance from environmentalists. Sumter Countians Organized to Protect the Environment (SCOPE), consisting largely of white professionals, business persons and landowners, organized early to protest what was perceived to be the indiscriminate handling of hazardous wastes and to call for greater accountability on the part of Chemical Waste Management and the regulatory agencies. Some years later, a more activist group known as Alabamians for a Clean Environment (ACE) was formed and called for the expulsion of Chem Waste and the closing of its facility altogether.

Aligned against this group are those who support Chem Waste, primarily because of the jobs and local revenue that this company provides to local government and civic groups. Chem Waste is the county's largest employer and provides local government and civic groups roughly $1.6 million each year in revenue, a major sum of money for a poor rural county. Consequently, in spite of environmental concerns many local residents at least tacitly support the presence of the industry.

Chem Waste is very effective in mobilizing this positive community support, buying newspaper ads and paying for special supplements in local newspapers to promote the image of a corporate good neighbor sincerely interested in the safety and well-being of local residents. Chem Waste also encourages its employees to become active in community groups, to which it contributes money and technical assistance, all of which is well publicized in the local press. One of the two local weekly newspapers in Sumter County refuses to accept for publication letters to the editor from members of ACE, claiming that they were seeking free publicity for their cause and should pay for their advertising the same as Chem Waste.

The economic benefits of Chem Waste operations in Sumter County are not shared equally by local residents. Chem Waste employs over 400 people, but not all who work at the facility are local residents of Sumter County. Indeed, the top paying technical and managerial jobs are held by people brought in by Chem Waste, something they felt necessary to do given the local unavailability of technically trained staff. Most of the revenues are appropriated to various organizations in the population centers of Livingston and York, both in the southern portion of the county. Smaller communities in the northernmost part of the county in which the landfill is located receive little of the revenue. Finally, some local landowners whose property is located near the facility have experienced a depreciation in the value of their property.

Not surprisingly, most of Sumter County's leaders are profoundly ambivalent to the presence of hazardous waste in their community, recognizing the jobs and tax revenues which Chem Waste operations provide, but also fearing the long-term consequences of these operations.

The fear of chemical contamination of groundwater, which is used by many residents in the area for drinking water, and the fear of accidental spills by trucks hauling wastes to the Chem Waste facility, are major concerns for many residents. It was these concerns that one respondent had in mind when he remarked that he was afraid Sumter County was "prostituting" itself, accepting short-term gain at the risk of long-term environmental risk.

The political environment of Sumter County makes emergency preparedness and planning extremely difficult for at least two reasons. First, preparedness of a magnitude equal to the threat posed by a major facility such as that found in Sumter County is extremely costly. We have earlier reported a critical need in the county for trained personnel in key emergency agencies in the county as well as more equipment and supplies capable of responding to a chemical emergencies (Faupel et al., 1987). Emergency officials are aware of this need, but as one official reports, "They're not willing to pay for it until after a disaster hits." Resistance on the part of local governments to underwrite emergency preparedness measures is understandable as it is difficult to justify expenditures for an event that might occur years in the future. More pressing needs are often perceived by the public.

A second source of controversy lies in the nature of emergency preparedness itself. If emergency planning is to be fully effective it must anticipate the worst possible scenario. Anything less could prove disastrous in an emergency situation. Planning for a worst case scenario chemical emergency serves to highlight an already controversial public issue. Preparing for a major chemical emergency is not only costly, which itself invites controversy, but it acknowledges that an emergency of major proportions is in fact *possible.* Moreover, preparedness is (or at least should be) a public process, involving educational strategies, public hearings, and disaster drills. All of these activities which are necessary for effective preparedness, are also periodic reminders of the hazardous technology hosted by the community.

The Merger of Public and Private Roles: Jurisdictional Dilemmas

Emergency response in the United States traditionally falls on the shoulders of local governments. While private enterprise clearly plays a role in major emergencies in the way of providing specialized equipment and even volunteer personnel, these roles are clearly understood to be subject to the authority of public agencies. Moreover, coordination of and preparedness for emergency response is recognized as a governmental responsibility, usually centered in the local Emergency Management Agency.

Public responsibility for emergency preparedness usually is justified on the grounds that major emergencies pose demands which are beyond the response capabilities of the private sector. Moreover, large scale emergencies affect entire communities which by definition makes emergency response a public concern. In the case of emergencies caused by natural disasters, because the locus of the agent is external to the social system, private responsibility for emergency response is a moot issue. Emergencies related to hazardous wastes also pose potentially serious consequences which affect entire communities. These are emergencies, however, for which responsibility can be more clearly assigned.

Most emergencies occurring at a hazardous waste facility are appropriately neutralized at the facility itself with company personnel and equipment. Our previous analysis of emergency preparedness in Sumter County found that Chem Waste was well prepared for emergencies which are contained within the site itself (Faupel et al., 1987). Less clear, however, is the jurisdictional authority for off-site emergencies and on-site incidents which impact beyond the facility itself. Technically, emergencies which pose a potential threat to the surrounding community fall within the jurisdictional domain of the public emergency response system. Federal law stipulates, however, that the facility itself is to determine whether an on-site incident needs to be reported to local officials. Moreover, in Sumter County local emergency officials indicated a high level of dependence on Chem Waste in handling emergency situations which may occur at any point in the county.

This dependence is a function of the limited expertise and resources available to local emergency personnel.[3] Chem Waste's willingness to provide such assistance is an example of this company's efforts to be a good corporate citizen. Chem Waste's support for local emergency planning efforts does, however, introduce a troubling question of possible conflicts between corporate and community interests in providing timely and complete information to local emergency officials and the public. In the politically charged atmosphere of Sumter County, it is possible that

[3]Personnel in Sumter County fire departments participate on a rotating basis in training courses at the "fire college" located in Tuscaloosa, where they receive training in handling chemical fires.

Chem Waste would publicly downplay the significance of both on- and off-site accidents involving hazardous wastes. The relative lack of local response capability weakens the ability of local emergency personnel to reach an independent judgement.

Local reliance on Chem Waste is not a solution to the problem of emergency preparedness for the county as a whole. The waste disposal facility is located in the northern part of the county, nearly 30 minutes away from the southernmost portion of the county. Many initial response activities, particularly the neutralizing of hazardous materials, require immediate action. Over-reliance on company personnel and equipment to accomplish these tasks may unnecessarily endanger the health and safety of emergency personnel and citizens in the area. At the same time, of course, when company resources are responding to an emergency in the community, emergency response capabilities at the plant site are proportionately reduced.

Bifurcation of Emergency Planning Phases and Responsibilities

The Federal Emergency Management Agency (1985) has identified four broad phases of emergency management, known collectively as the Integrated Emergency Management System (IEMS). These phases are mitigation, preparation, response, and recovery (Clary, 1985; McLoughlin, 1985; Petak, 1985; Zimmerman, 1985). Mitigation involves all of those considerations which reduce the likelihood of an emergency situation occurring in the first place. Mitigation measures related to the handling of hazardous wastes are carefully detailed in EPA regulations and permits and should be integrated into the standard operating procedures of the hazardous waste industry. Preparation entails such considerations as warning, stockpiling of relevant supplies and equipment, etc. Response activities, by contrast, include search and rescue, evacuation and, in the case of waste related emergencies, containing toxic releases and extinguishing fires. Recovery involves those extended activities necessary to bring the community back to a state of normalcy, including provision for long-term financial assistance and, particularly relevant to chemical hazards, the existence of sufficient liability insurance on the part of the chemical industry.

Until recently, however, emergency planning was understood to consist of preparedness and response measures as reflected by the social science literature on disaster (Mileti et al., 1975). Mitigation and recovery activities have been taken up largely by urban planners (Baker, 1977; Burby and French, 1981; County of Santa Clara Planning Department, 1977; David and Mayer, 1984; Francaviglia, 1978; Galloway and Huelster, 1971; Platt, 1976, 1978, 1982; State of Alaska, 1981; Willis, 1976). This bifurcation is mirrored in the division of labor among responding agencies. Traditional emergency response agencies such as police and fire departments, hospitals, ambulance services, and emergency management agencies (Civil Defense) have been mandated the roles of preparedness and response. Mitigation strategies have been relegated to regulatory agencies, planning offices, and zoning boards. Similarly, long-term recovery has typically been understood to be the domain of federal and state governments and has not been well integrated into the local emergency planning process.

This situation is particularly troublesome for hazardous waste emergency planning. As observed at the outset, hazardous wastes pose both *chronic* and *acute* threats. These two types of threats are closely interrelated. Groundwater and aquifer contamination, for example, while usually considered chronic threats, become acute threats when water supplies become so contaminated as to pollute a community's water supply. Similarly, the chronic threat of exposure of workers to hazardous material becomes acute in the face of major equipment malfunction where numbers of workers are employed. The hazardous waste industry by definition poses the threat of chronic environmental and public health problems which may suddenly turn into serious acute threats necessitating emergency preparedness and response by local agencies.

The close interrelationship between mitigation, preparedness, response, and recovery is not, of course, unique to hazardous wastes. The need for an integrated emergency management system has been recognized for a number of years (Kemp, 1984; Lewis et al., 1977; O'Keefe and Westgate, 1977; Rubin, 1979, 1982, 1983; Wenger et al., 1985; Whitaker, 1978). Only recently, however, have practical steps been taken to establish such integrated systems. The establishment of the Integrated Emergency Management System by FEMA represents a major attempt to structurally integrate these emergency phases at the local community level. Similarly, Title III of the Superfund Amendments and Reauthorization Act of 1986 (aimed primarily at manufacturers using, producing, or storing hazardous materials and not specifically at hazardous wastes) was passed to provide local communities with the information necessary to develop integrated emergency management systems.

FEMA requires that local communities address all of these phases in formal disaster planning if they are to receive federal funds for emergency preparedness. The Sumter County Emergency Management Agency plan has integrated each of the four phases in its community plan. In spite of this, however, the emergency management system in Sumter County remains bifurcated. Agencies responsible for mitigation and recovery measures (planning office, zoning board, regulatory agencies, federal loan agencies, etc.) are scarcely mentioned in the written plan at all and play no role whatsoever in the on-going emergency planning process. This situation, which is a vestige of the historic bifurcation of "urban planning" vs. "disaster planning," renders emergency management less than comprehensive and inhibits successful coordination of the emergency planning process.

CONCLUSIONS

Chemical hazards possess unique features which are not shared by natural disasters. Unlike most natural disasters, for example, appropriate response to chemical emergencies is not always clear. Chemical properties, such as flash points, toxicity, and vapor density, are highly variable. Hence, containment measures which may be effective with one chemical may only exacerbate an emergency involving a different substance. Moreover, emergencies involving hazardous wastes pose the special problem of unpredictable synergism caused when several impure chemical compounds are accidentally mixed, making response even more difficult.

These unique characteristics of chemical hazards require special considerations in response planning which extend beyond the current capacity of traditional emergency preparedness agencies. First, *specialized equipment and supplies* are required. Special foams are required for extinguishing many chemical fires, as are gas masks and similar protective devices. Second, trained personnel are needed to adequately respond to chemical emergencies. Finally, local residents and community leaders must recognize the *possibility* of an emergency and devote necessary energy and financial resources to planning and preparing for a serious accident before such an event occurs. This is particularly difficult in a social context where emergency planning is highly politicized, as is the case in Sumter County.

Expenditure of public monies for emergency preparedness will attract increased public attention to the larger issues of hazardous wastes. Money will not solve all the problems associated with emergency preparedness, but additional funds must be made available if Sumter County residents are to establish an independent planning and response capacity. At present, the county government receives roughly $1.6 million each year in revenue from a $7.00 per ton fee charged for all wastes entering the Emelle facility. This revenue is divided among many government and civic groups within Sumter County according to a formula set by elected leaders representing the county in the state legislature. These funds are used to meet many important local needs. Consideration should be given to the question of whether adequate support is being provided for local preparedness for handling an emergency involving hazardous wastes.

REFERENCES

1. Baker, Earl J.: Public attitudes toward hazard zone controls. *Journal of the American Institute of Planners* (October):401–408, 1977.
2. Bucher, Rue: Blame and hostility in disaster. *American Journal of Sociology* 62: 467–475, 1957.
3. Burby, Raymond and French, Steven P.: Coping with floods: the land use management paradox. *Journal of the American Planning Association* (July): 289–301, 1981.
4. Clary, Bruce B.: The evolution and structure of natural hazards policy. *Public Administration Review* 45:20–28, 1985.
5. County of Santa Clara Planning Department: Considering seismic safety in land use planning. *Ekistics* 260:38–45, 1977.
6. David, Elizabeth and Mayer, Judith: Comparing costs of alternative flood hazard mitigation plans: the case of Soldiers Grove, Wisconsin. *Journal of the American Planning Association* 50:22–35, 1984.
7. Drabek, Thomas E.: Disaster in aisle 13. Columbus, Ohio: College of Administrative Science, The Ohio State University, 1968.
8. Drabek, Thomas E. and Quarantelli, E.L.: Scapegoats, villains and disasters. *Transaction* 6:33–38, 1967.
9. Dynes, Russell R., Quarantelli, E.L. and Kreps, Gary A.: *A Perspective on Disaster Planning.* Disaster Research Center Report Series #11. Columbus, Ohio: Disaster Research Center, The Ohio State University, 1981.
10. Environmental Protection Agency: *The Hazardous Waste System.* Office of Solid Waste and Emergency Response. June 1987. Washington, D.C.: Environmental Protection Agency, 1987.
11. Faupel, Charles E.: *The Ecology of Disaster: An Application of a Conceptual Model.* New York: Irvington, 1985.
12. Faupel, Charles E., Bailey, Conner and Williams, Marcus: *Hazardous waste and*

emergency planning: a case study of Sumter County, Alabama. Bulletin #587, Alabama Agricultural Experiment Station, Auburn, Alabama, 1987.

13. Federal Emergency Management Agency: *Guide for Development of State and Local Emergency Operations Plans.* Washington, D.C.: U.S. Government Printing Office, 1985.
14. Francaviglia, Richard V.: Xenia Rebuilds: effects of pre-disaster conditioning on post-disaster redevelopment. *Journal of the American Institute of Planners* (January):42–51, 1978.
15. Galloway, Thomas D. and Huelster, Ronald J.: Planning literature and the environmental crisis: a content analysis. *Journal of the American Institute of Planners* (July):269–273, 1971.
16. Kemp, Roger L.: The city manager's role in emergency management. *Public Management* 66:9–12, 1984.
17. Levine, Adeline Gordon: *Love Canal: Science, Politics and People.* Lexington, Mass.: Lexington Books, 1982.
18. Lewis, James, O'Keefe, Philip and Westgate, Kenneth N.: A philosophy of precautionary planning. *Mass Emergencies* 2:95–104, 1977.
19. McLoughlin, David: A framework for integrated emergency management. *Public Administration Review* 45:165–172, 1985.
20. Mileti, Dennis, Drabek, Thomas and Haas, Eugene: *Human Systems in Extreme Environments: A Sociological Perspective.* Program on Technology, Environment and Man, Monograph #21. Boulder, Colorado: University of Colorado, 1975.
21. Neal, David M.: Blame assignment in a diffuse disaster situation: a case example of an emergent citizen group. *International Journal of Mass Emergencies and Disasters* 2:251–266, 1984.
22. Nigg, Joanne M. and Cuthbertson, Beverly Ann: Pesticide applications near urban areas: a "crisis in confidence" for public health agencies. *Journal of Health and Human Resources Administration* 4:284–301, 1982.
23. O'Keefe, Phil and Westgate, Ken: Preventive planning for disasters. *Long Range Planning* 10:25–29, 1977.
24. Perrow, Charles: *Normal Accidents: Living With High Risk Technologies.* New York: Basic Books, 1984.
25. Petak, William J.: Emergency management: a challenge for public administration. *Public Administration Review* 45:3–7, 1985.
26. Platt, Rutherford H.: The National Flood Insurance Program: some midstream perspectives. *Journal of the American Institute of Planners* (July):303–313, 1976.
27. Platt, Rutherford H.: Coastal hazards and national policy: a jury-rig approach." *Journal of the American Institute of Planners* (April):170–180, 1978.
28. Platt Rutherford H.: The Jackson flood of 1979: a public policy disaster. *Journal of the American Planning Association* (Spring):219–231, 1982.
29. Quarantelli, E.L.: *Sociobehavioral Responses to Chemical Hazards: Preparation for and Responses to Acute Chemical Emergencies at the Local Community Level.* DRC Book and Monograph Series #17. Newark, Del.: Disaster Research Center, University of Delaware, 1984.

30. Rubin, Claire B.: *Natural Disaster Recovery Planning for Local Public Officials.* Columbus, Ohio: Academy for Contemporary Planning, 1979.
31. Rubin, Claire B.: Managing recovery from a natural disaster. *Management Information Service Report* 14:1–14, 1982.
32. Rubin, Claire B.: Recovery from natural disaster: a larger role for the fire chief. *Fire Chief* 27, 1983.
33. State of Alaska: *Floodplain Management for Alaskan Communities.* Anchorage: Department of Community and Regional Affairs, 1981.
34. United States Bureau of the Census: *County and City Data Book.* Washington, D.C.: U.S. Government Printing Office, 1983.
35. Veltford, Helene and Lee, George F.: The Cocoanut Grove fire: a study in scapegoating." *Journal of Abnormal and Social Psychology* 38:138–154, 1943.
36. Walsh, Edward J.: Local community vs. national industry: the TMI and Santa Barbara protests compared. *International Journal of Mass Emergencies and Disasters* 2:147–163, 1984.
37. Wenger, Dennis E., James, Thomas F. and Faupel, Charles E.: *Disaster Beliefs and Emergency Planning.* New York: Irvington, 1985.
38. Whitaker, Hilary: *State Comprehensive Emergency Management: Final Report of the Emergency Preparedness Project.* Center for Policy Research, National Governer's Association. Washington, D.C.: U.S. Government Printing Office, 1978.
39. Willis, J.D.: *Flood Insurance—Asset or Liability in Land Use Control.* Research Report 76:2. Ottowa: Emergency Planning Canada, 1976.
40. Zimmerman, Rae: The relationship of emergency management to governmental policies on man-made technological disasters. *Public Administration Review* 45: 29–39, 1985.

Chapter 9

PSYCHOSOCIAL IMPACTS ON TRIAL: THE CASE OF HAZARDOUS WASTE DISPOSAL[1]

MICHAEL R. EDELSTEIN

INTRODUCTION

In courtrooms, administrative hearings, and in the field of social impact assessment, a professional paradigm is taking form for the definition and identification of psychosocial and community impacts related to the siting, operation, and aftermath of hazardous waste disposal. Because much of this development is occurring in the applied consulting context rather than as part of traditional academic research, discourse over definitions, epistemology and methodology has been largely constrained to adversarial peer review within hearings and trials or to the bureaucratic realm of environmental impact review. The result is that social science theory, methodology, and epistemology have been put on trial. Unlike the relative tolerance of the scientific community for multiple theories and approaches, here there may be black-and-white determinations: one approach wins, the other loses. As a result, key decisions are being framed by attorneys and made by judges in both civil and administrative courts that define the form and content of the field of psychosocial impact assessment relating to hazardous waste siting.

In this chapter, litigation of psychosocial impacts is briefly reviewed and analyzed. Then a case study is presented that illustrates the profession-defining issues involved in the adjudication of psychosocial impacts.

[1]The author's testimony and research for the case study discussed in this chapter was prepared at the request of and underwritten by funds received from the New York law firm of Milbank, Tweed, Hadley & McCloy. The author wishes to thank Deborah Kleese, Ph.D.; J. Stephen Kroll-Smith, Ph.D., and Brent K. Olsson, esq. for their comments on a draft of this article and Adeline G. Levine, Ph.D. and Murray Levine, Ph.D. for their helpful discussion of the case. I am also grateful to Dr. Kroll-Smith for his assistance in conducting the reported research.

Finally, conclusions are drawn for the study of these impacts in the context of hazardous waste disposal.

LITIGATING PSYCHOSOCIAL IMPACTS OF HAZARDOUS FACILITIES

The courtroom has been an important context for defining the psychosocial impacts of hazardous facilities, much as it has spurred the psychological examination of other current problems. By way of illustration, two prominent examples will be reviewed.

An important 1983 U.S. Supreme Court decision involved the claim by the citizens group People Against Nuclear Energy (PANE) that, under the 1969 National Environmental Policy Act (NEPA), the Nuclear Regulatory Commission should have reviewed psychological harm in its decision to restart the Three Mile Island 1 nuclear reactor. PANE argued that the operating reactor would provide an ever present reminder of the 1979 accident at the Three Mile Island 2 nuclear reactor, causing severe psychological damage to people living nearby the facilities, severely damaging the stability, cohesiveness and well-being of nearby communities, and creating stigma for the area. In response, the Nuclear Regulatory Commission argued that PANE's case was not based on an analysis of actual physical danger (e.g., an expert risk assessment), but rather upon public perceptions of dangers (Marshall, 1982).

On May 14, 1982, the U.S. Appeals Court for the District of Columbia found for PANE, confirming that NEPA included a requirement to consider psychological health. The majority wrote that "... effects on psychological health are effects on the health of human beings." The court distinguished between what they called "true psychological stress" and ordinary "socioeconomic anxieties." By psychological stress, they referred to posttraumatic anxieties and physical effects due to "fears of recurring catastrophe," as found at Three Mile Island. Socioeconomic anxieties were illustrated by the feelings of middle-class homeowners when a housing project is located nearby. The dissent by Judge Wilkey argued that the decision would now hinge, not on an assessment of the risk, but on "how much fear is worked up, from whatever source" (Marshall, 1982, p. 1204).

The decision was appealed to the U.S. Supreme Court by the Nuclear Regulatory Commission and Metropolitan Edison, the utility that owned the Three Mile Island plants. The Nuclear Regulatory Commission argued that the consideration of psychological effects is not demanded under NEPA unless caused by a physical effect, which was not the case. The utility argued that psychological harm cannot be measured, and, therefore, cannot be considered (Jordan, 1984). An Amicus Curiae filed on behalf of PANE by the American Psychological Association on September 30, 1982 cited studies of the psychological effects of the Three Mile Island 2 accident as a factual record for proximate psychological effects from restarting the Three Mile Island 1 nuclear reactor. In a second brief, the American Psychological Association further argued that Congress intended that NEPA cover psychological impacts. Short of studying direct injury to the brain, the American Psychological Association argued that psychological impacts can only be studied by measuring perceptions. Thus, the court could not rule out the measurement of perceptions (American Psychological Association, 1982).

The Supreme Court decision, written by Justice Rehnquist for a unanimous court, determined that NEPA was intended to address only effects on the physical environment. Risk and the perception of risk were seen as too far removed in a causal chain beginning with the physical action of restarting the reactor:

> But a risk of an accident is not an effect on the physical environment. A risk is, by definition, unrealized in the physical world" (U. S. Supreme Court, p. 4373).

Furthermore, the Three Mile Island 2 accident was viewed as irrelevant to the consideration of the restart of the Three Mile Island 1 nuclear reactor because it was a different action and because it occurred in the past; NEPA, the court ruled, addresses the future.

In these ways, the Court limited the consideration of psychological impacts. The influence of the decision can be seen, for example, in the guidelines issued by the Department of Energy for reviewing impacts from proposed high level nuclear waste repository sites (*Federal Register*, December 6, 1984, p. 47747):

> Perceived risk . . . is not an appropriate topic for general repository-siting guidelines; it is a subjective condition that cannot be fairly compared among sites.

However, the Court did not exclude psychological impacts from NEPA review. In his concurring opinion, Justice Brennan clarified that psychological impacts must be assessed under NEPA when they stem from the "direct sensory impact of a change in the physical environment," but not due to the "perception of risk" (U. S. Supreme Court, 4375). Thus, a clear causal link between psychological impacts and an event can be established when psychosocial consequences are seen as secondary or indirect impacts of an action, rather than merely as risks of a possible accident (Jordan, 1984; also Hartsough and Savitsky, 1984; Sorenson et al., 1987; Hartsough, 1989).

In sum, the courts used the PANE litigation to legitimate the assessment of psychological impacts under NEPA, but they set forth a series of tests to assure that frivolous claims of psychological injury did not result; namely, the consideration of psychological impact required that a change to the physical environment result in a degree of reliably measured stress severe enough to threaten physical health. The PANE case was lost because it did not adequately meet these tests (Hartsough and Savitsky, 1984).

A second context for the development of psychosocial impact study has been toxic torts. An early and influential toxic tort decision was Ayers v. Jackson Township (Edelstein, 1988). This suit, brought by 339 residents of the Legler section of Jackson Township, New Jersey, resulted from contamination by the municipal landfill of the groundwater used by residents. Evidence for the plaintiffs included studies of both the psychosocial impacts (Edelstein, 1982) and psychological damage (Gibbs, 1982) to the plaintiffs due to the contamination. Experts testifying for the defendents offered no direct research about psychological impacts, instead challenging the methodology used by the plaintiff's consultants.

In 1983 a New Jersey jury held the township responsible for the contamination and awarded $15 million dollars to the plaintiffs. The award included $2 million for emotional distress and $5.4 million for diminished quality of life. The decision was appealed twice, with the New Jersey Supreme Court finally determining in 1987 that the quality of life portion could stand, but not the emotional damage award. The disallowance of this award reflected a technicality in New Jersey law prohibiting the granting of "pain and suffering" damages against municipalities. Unlike subjective factors relating to "pain and suffering," the court cited nuisance law in arguing that life adjustments due to the contamination of the groundwater

> objectively affect the quality of life by causing an interference with the use of one's land through inconvenience and the disruption of daily activities (New Jersey Supreme Court, A-83-84, decision in *Ayers v. Jackson Township,* May 7, 1987, Judge Stein citing in part from 202 *N.J. Super.* at 118).

Thus, overall, the Ayers suit provides a precedent recognizing psychosocial damages due to toxic exposure, both in the sense of impacts on quality of life and for psychological damage.

OPPOSING PERSPECTIVES FOR VIEWING PSYCHOSOCIAL IMPACTS

The PANE and Ayers decisions illustrate how litigation can contribute to the definition of the emerging field of psychosocial impact review relating to hazardous facilities. The Ayers case involved the assessment of psychosocial impacts that had already occurred, the PANE case sought to consider future impacts of a proposed event. While the PANE argument was inherently broad, in keeping with prior precedent-setting litigation involving the Buffalo Creek disaster that had recognized various individual and community impacts (see Erikson, 1976), the effect of the PANE litigation was the narrowing of consideration of psychological impacts to the stress construct (Hartsough, 1989). In contrast, the pain and suffering and nuisance arguments in the Ayers suit were made and decided more broadly. The Ayers plaintiffs prevailed, although the emotional damage award was overturned on a technicality; in the PANE litigation, the argument for psychosocial impacts lost.

The arguments and decisions in the two cases are ambivalent in that they both include expanding and limiting themes for the definition of psychosocial impact assessment. By way of illustration, Table 9-1 lists key themes from both law suits. As shown, an **expanding perspective** on psychological impact views perceptions as inherently linked to physical conditions. As secondary or indirect impacts of direct sensory impacts, perceptions are seen as a legitimate means for studying the relationships between events and outcomes as a basis for predicting future psychosocial impacts. Consideration of impacts is not restricted to the stress construct, to individual impacts, nor to a given methodological orientation. Impacts on quality of life are recognized as are more severe mental health effects.

The contradictory assumptions involved in the **expanding** and **limiting** models for defining psychological impact suggest their utility for examining expert testimony on this issue. Suggested is that, in the adversarial context of litigation, there is a predictable divide between experts adhering to these differing models.

Table 9-1.
Some Key Elements of Expanding and Limiting Models for Defining Psychosocial Impacts of Hazardous Sites

Expanding Model	*Source*
The reminder of the accident created by restart of TMI 1 would damage the stability, cohesion and well-being of the nearby communities and create stigma.	PANE
The effects from the accident can serve as a basis for predicting the effects of the restart.	APA
Psychological effects can only by studied by measuring perceptions.	APA
Effects on psychological health are effects of health.	U.S. Appeals Court
Psychological impacts must be assessed under NEPA when they stem from a direct sensory impact or change in the physical environment.	Justice Brennan
Psychosocial impacts can include both pain and suffering and nuisance claims.	New Jersey Supreme Court
Limiting Model	
Public perceptions are not the same as physical danger.	NRC
Psychological harm cannot be measured.	Met. Ed.
True psychological stress differs from socioeconomic anxiety.	U.S. Appeals Court
Fear can be worked up from any source.	Justice Wilkey
A risk is unrealized in the physical world.	U.S. Supreme Court.

In contrast, the **limiting perspective** differentiates and separates subjective perceptions from objective physical conditions. Fears are recognized only when "true psychological stress" is identified, suggesting that there is a threshold for recognizable impact. Perceptions are associated with intangible future risks, not with actual physical events. Prediction of future psychological impacts is limited by the requirements of controlled stress research.

Based upon the Ayers case, these models can also be used to predict the approach taken by the respective experts in approaching psychosocial impacts. Thus, experts having a **limiting perspective** can be expected to take a quantitative orientation to the controlled study of the stress construct, making expert arguments about the best methodological approach, but without collecting primary data to identify impacts. Because of its more limiting and conservative implications, this approach fits well with the needs of applicants in siting cases and defendants in toxic tort litigation because it limits the discovery of impacts that might hurt the application or defense.

In contrast, experts having an **expanding perspective** will evidence a wider operational definition and approach to the measurement of psychological impact, often collecting data in the specific context to identify impacts. Because the expanding approach provides substantive proof of impacts, if present, it best serves opponents of toxic sites and plaintiffs in litigation.

The next section of the chapter involves the presentation of a case study of an adversarial proceedings dealing with the siting of a hazardous facility. The implications of the case study for the further definition of psychosocial impact assessment will then be analyzed.

PSYCHOSOCIAL IMPACTS IN THE CECOS SITING HEARINGS

The CECOS waste disposal complex is located on the eastern boundary of the City of Niagara Falls, New York, lying largely in the Town of Niagara. To the west is a large industrial belt. Also bounding the site are commercial strips, major roads and highways, and several residential areas, housing thousands of people. Nearing completion of its fifth hazardous waste landfill in 1985, CECOS, a division of Browning Ferris, Inc., applied to the New York State Department of Environmental Conservation (DEC) for permission to build a sixth facility at the site, to be called the Secure Chemical Residue Facility (SCRF #6). However, prior to the start of permit hearings before an administrative law judge (ALJ), the New York State legislature mandated in 1987 that hazardous waste sites also be approved by a special siting board consisting of three local citizens from the candidate site area and representatives of five state departmental commissioners. While the ALJ would write a decision based upon hazardous waste disposal laws and the New York State Environmental Quality Review Act, the siting board would decide if the project was in the public interest.

For these hearings, interested party (intervener) status was granted to six local environmental and citizen organizations known collectively as the Concerned Citizens Organization (Concerned Citizens).[2] In addition to the regional staff from the DEC and CECOS, the other parties to the proceedings were the City of Niagara Falls, the County of Niagara and the Niagara County Board of Health, as well as the Town of Niagara, the Province of Ontario and the New York State Consumer Protection Board.

PSYCHOLOGICAL IMPACTS AS AN ISSUE FOR ADJUDICATION

Among the issues that the Concerned Citizens sought to have adjudicated were the psychological impacts of the SCRF #6 proposal. However, while such social impacts must be considered under the New York State Environmental Quality Review Act, the ALJ found that the intervener's case for psychological impacts failed to trigger the criterion for a substantive and significant issue, namely the potential for resulting in permit denial, major permit modification or the imposition of significant permit conditions (Pearlstein, Ruling of May 20, 1988). The judge was willing to receive written comment on psychological impacts, but there was to be no testimony by experts on the issue.

The interveners appealed the ruling to the DEC Commissioner and the Siting Board, citing the Metropolitan Edison v. PANE precedent that risk of harm from a hazardous facility can have a psychological impact. They argued that there are direct psychological impacts in the SCRF #6 case and, specifically, that the

> *subjective* concerns of the citizens . . . are relevant because, objectively justifiable or not, it is these concerns which will interfere with the everyday lives of the citizens, and hence which will have an impact upon the public interest (Concerned Citizens Organizations, 1988, p. 14).

In their interim decision granting the appeal, the siting board recognized that

---

[2]Six organizations belonged to this consolidated group: the Ecumenical Task Force of the Niagara Frontier, Campaign to Save Niagara, Great Lakes United, LaSalle and Niagara Demand, Society to Oppose Pollution in Towns, and Evershed Restoration Association. Party status reflects the judgment of the ALJ that these groups represented people having a legitimate interest in the outcome of the hearings. As an intervener, Concerned Citizens were able to address the issues selected for joint adjudication by the ALJ and siting board, call witnesses and cross-examine the witnesses called by other parties. With the assistance of New York Governor Cuomo, the Concerned Citizens received pro bono representation by the law firm of Milbank, Tweed, Hadley & McCloy.

> it is clear that many community residents find the construction of SCRF #6 to be distressing, [but that at the time of the issues hearing] no offer of proof was made beyond the statements of the individual residents themselves (Industrial Waste Siting Board, 1988, p. 3.)

Defining psychological impacts as relevant to their determination of whether the project was in the public interest, the siting board granted the interveners an opportunity to present the proposed testimony of an expert witness to offer "proof" that would advance the record beyond what the Siting Board had referred to as the "subjective state of mind" of residents who had given testimony at the public hearing (Industrial Waste Siting Board, 1988, p. 3). The Siting Board's decision was accepted by the Commissioner of the DEC (Jorling, 1988). As a result, psychological impacts became an issue for adjudication in the SCRF #6 hearings.

In the Realm of Experts

The Siting Board Decision thus raised a theme also found in the Metropolitan Edison V. PANE deliberations, that psychological impacts belong in the realm of experts. The implication is an important one, namely, that citizens who address their own concerns at hearings cannot prove their claims with nonexpert evidence alone. Their burden of proof can only be sustained by evidence presented by an expert. In other terms relevant to this analysis, citizen claims remain subjective until proven to be objective by an expert testifying under oath (see Edelstein, 1987).

In the SCRF #6 hearing, four experts were called to testify on the issue of psychological impacts. CECOS called one principal expert, Vincent Covello of Columbia University and one rebutal witness, Robert Gatchel of the University of Texas. The interveners called the author as their principal witness. A second expert, Gerald Goldhaber of Goldhaber Research Associates, was called to discuss technical issues related to a survey study done for the interveners. In all, the presentation of testimony on psychological impacts, including cross-examination, lasted for five days in early November, 1988. Both sides offered prefiled testimony prior to the hearings, as well as final briefs and response briefs subsequent to testimony.[3,4]

---

[3]The parties opposing the SCRF #6 permit divided up responsibility for the many issues adjudicated during months of hearings. Concerned Citizens were the lead intervener with regard to psychological impacts.

[4]Cavello's testimony is found in Cavello (1988); also transcripts of the SCRF #6 hearing, Volumes 30 and 31. Gatchel's testimony is found in Volume 34 of the hearing record. Edelstein's testimony is found in Edelstein (1988b) and in Volumes 32 and 33 of the hearing transcripts.

Some of the key differences in the testimony regarding psychological impacts is reflected in the training and backgrounds of the experts. As their primary witness, CECOS hired a sociologist noted for his work in risk perception and experienced in administering large-scale, controlled research projects. Their rebuttal expert was a clinical psychologist involved in controlled, quantitative studies of psychophysiological stress, as well as in the practice of therapeutic behavior modification. In contrast, the intervener selected as their expert a qualitatively-oriented environmental/ social psychologist with experience in social impact assessment.

A further difference was the extent to which the experts depended upon theoretical arguments and/or field investigation. The experts for CECOS collected no field data. Covello indicated that his testimony was based upon a review of documents and a portion of a file of newsclippings providing by CECOS, a tour of CECOS and the neighborhood guided by CECOS personnel and discussions with CECOS personnel and consultants during a day-long visit to the site. Gatchell only reviewed documents.

In contrast, the intervener's expert combined theory with observations from a tour and several days spent in the CECOS area, and an analysis of data from three convergent research studies conducted in the two months before the hearing (Edelstein, 1988b). These studies are summarized below.

Group Interviews

In September 1988, six focused group interviews were conducted with a total of 30 community residents in order to obtain an indepth understanding of the ways that the existing CECOS operations were affecting residents and to explore any concerns that they had regarding the SCRF #6 application. The interviews were held at the 66th Street School, located within view of the CECOS site. Five of the interviews involved a sample of residents from nearby the CECOS site recruited mostly through random phone calls. A sixth group involved key informants, leaders or active members of the intervening citizens groups. The sample involved a cross section of the community in terms of activism and concern, employment, income, location, age, and sex.

Documentary Research

Beyond the interviews, project documents, census data, and government files were reviewed. An additional analysis was made of a file consisting of some 250 newspaper clippings provided by the Ecumenical

Task Force. The news file provided a basic chronology of events regarding the CECOS site spanning the years 1985 to 1988 and reflecting two independent newspapers. Second, the file provided a means for confirming or disconfirming statements made by the interview participants. Finally, the file was instructive regarding the degree that media may have affected public perceptions of CECOS and the SCRF #6 application.

The Community Survey

Finally, the firm of Goldhaber Research Associates (GRA) was hired to survey a proportionate 10 percent sample randomly drawn from reverse telephone directories listing 2,800 residents in three target neighborhoods downwind and nearby CECOS that were selected after a review of records of community complaint about CECOS (Hopkins, August 31, 1988). Using a 35-item survey instrument, GRA conducted telephone interviews with 280 respondents from the large, single-family blue collar LaSalle area, a trailer park called Sabre Park tucked into the southeast edge of the CECOS site, and a middle-class, single family home section on Third Avenue, located perpendicular to the CECOS site.

The conclusions drawn by the experts for the relative parties are summarized below.

Summary of Applicant's Conclusions

In their direct and rebuttal testimony on psychological impacts, CECOS' experts made the following points:

1. That psychological impacts can only be identified through the use of standardized, quantitative measures of stress symptomology, most notably the use of the SCL-90R and PERI instruments, preferably with prepost or control-group comparisons.
2. That the conditions for ideal controlled research were absent from the SCRF #6 situation, and, therefore, that little could actually be established about SCRF #6 as the cause of stress symptoms.
3. That it was not surprising that residents of the area around CECOS opposed SCRF #6 given the characteristics upon which risk is normally perceived and the presence of biased media coverage. Perceived risk was not directly related to stress, however.
4. Because SCRF #6 represents a state-of-the-art hazardous waste landfill, it actually entails little risk. Therefore, community concerns do not accurately reflect reality. As a result, psychological impacts related to

SCRF #6 can be reduced and mitigated through educational programs and other stress-management techniques.

Summary of the Interveners' Conclusions

Based upon the above research, the interveners's expert drew a series of conclusions summarized below.

1. Psychological impacts were defined inclusively, as follows (Edelstein, 1988b):

> Psychological impacts involve the effects of an event or a series of events on the behavior, the cognitions and the emotions of people exposed to the event(s) (p. 6).

Furthermore, the terms "psychosocial impact" and "psychological impact" were used interchangeably because

> . . . at bottom, psychological well-being reflects psychosocial factors. . . . Individuals never behave in isolation. Their actions are moderated by—and in turn affect—their social environment (p. 9).

Finally, there is a time dimension to these impacts.

> Psychological or psychosocial impact assessment involves the projection of future impacts on the well-being of people—individually and collectively—that can reasonably be expected to occur due to an event or series of events, such as the planning, permitting, construction, operation, closure and post-closure phase of a hazardous waste landfill such as SCRF #6 (pp. 12–23).

2. The task of identifying potential psychological impacts of the SCRF #6 project was approached within the context of impact assessment, where the prediction of future impacts from a proposed facility is required. To address these problems in the SCRF #6 studies, two approaches were combined. The first derives from the view that predictions of future psychosocial impacts must be grounded in a firm understanding of present social process, since the future evolves, at least in part, from the present dynamics of the community (see Edelstein, 1978; Burdge, Field and Wells, 1988). Second, because an existing hazardous waste facility was operating at the CECOS site, it is possible to argue that impacts and concerns related to the existing operation are a legitimate basis for projecting impacts from a future facility involving comparable operations. This comparison holds constant all factors other than time that relate to the community, as well as most factors relating to the facility (Edelstein,

1980). Thus, data collected regarding impacts of the current CECOS operation combined with an understanding of the historical process of the community serve as a basis for judgment about future effects.

3. The large proximate population to the CECOS site places thousands of people at risk for psychosocial impacts related to the SCRF #6 facility. Psychological uncertainty will cause effects even in the case where the site operates as planned. These impacts reflect the influence of the community's history, environment, and social process. In particular, the Love Canal incident and the subsequent discovery of scores of additional hazardous waste sites in the vicinity has had both direct and indirect effects upon the area. As a consequence of this unique toxic legacy, residents of Niagara Falls may be the most sensitized population anywhere to the kinds of impacts threatened by SCRF #6. This sensitivity is evident in the local near-consensus in opposition to the SCRF #6 facility, as indicated by the coalition of governments opposing the application and by expressed opposition to SCRF #6 by 88 percent of the survey respondents.

4. Many residents are subjected to a cumulative impact from direct stressors stemming from CECOS and other nearby industrial facilities. For example, some respondents reported that they had to limit outdoor activities and keep their windows shut because of noxious odors.

5. CECOS has also forced residents to make lifescape adjustments (see Edelstein, 1988), shifting fundamental beliefs about their lives. Specifically:

a. Residents expected that they or their families would experience CECOS-related health problems.

b. There was a marked loss of perceived individual and community control linked, in part, to the inability of residents to prevent the prior CECOS permits. Opposition to SCRF #6 appeared to be a last attempt to regain control over their community, making the SCRF #6 permit a watershed decision;

c. There has been an alteration in perception of the Niagara Falls environment since the late 1970s. Previous community images involved the scenic wonder of the falls, the peacefulness of the residential neighborhoods, and the prosperity of the industrial center. After Love Canal, a toxic landscape emerged to dominate these other images. The Residents clearly saw SCRF #6 as an expansion of this toxic landscape;

d. There was also indication of the inversion of home in the surrounding communities. Inversion of home involves the loss of a sense of security and identity in the home; the home incurs a perceived loss of both psychological and financial value. Some residents had reportedly relocated from the area due to their concerns regarding CECOS. Others were uncertain about whether to invest in upkeep and renovations for their homes. There was a general perception that CECOS had harmed the value and saleability of nearby residences. With nearly a quarter of the residents sampled in the survey former residents of the Love Canal neighborhood, residents were also sensitive to the possibility of having to relocate;

e. A final lifescape shift already evident in the community was distrust of state and local officials, as well as distrust of CECOS. Reflecting their experience of previous waste facilities touted as state-of-the-art, residents additionally expressed a general lack of faith in the SCRF #6 design, despite its use of current technology. While few residents perceived benefits due to the SCRF #6 proposal, most saw it as the source of future costs resulting from a failure of technology and resulting contamination.

6. In examining the SCRF #6 project, numerous direct and indirect stressors were identified, varying in their impacts over different phases of the project. About half the respondents to the survey reported that the CECOS site has been a source of stress. Direct stressors, significant, not merely as noxious conditions, but because of their meaning to residents, included visual impacts, odor, dust, soot, and traffic.

While SCRF #6 itself would not be particularly visible outside the property, "Mt. CECOS," as the large sanitary landfills rimming the perimeter of the CECOS site were called locally, symbolized a hidden threat of toxic release from the hazardous waste facilities at the site. These landfills are visible from the main highways in Niagara Falls and, given the otherwise flat topography, from much of the surrounding community.

7. There was also strong indication of anticipatory fear regarding CECOS. More than 90 percent of the survey respondents were familiar with the site and evidenced related concerns. Seventy percent indicated that there could be future harm associated with the facility and 73 percent that CECOS is a threat to their well-being. Only 15 percent of the respondents believed the likelihood of an on-site accident at CECOS to be extremely slight and 19 percent a transportation-related accident to be extremely slight. Eighty six percent believed that there is a high potential for water contamination.

Analysis of press coverage for threat messages that would induce a fear of the site (e.g., discussions of how chemicals can cause cancer) revealed relatively few of them. Therefore, it was concluded that media bias did not create the anticipatory fear.

8. There was also evidence of environmental stigma related to the CECOS site and to Niagara Falls' toxic landscape in general. Residents complained that the toxic landscape had superceded "The Falls" as a major tourist attraction. Canadians involved in the tourist trade feared that their city would be similarly stigmatized by a proposed Canadian hazardous waste site (Pannell, Kerr, Forster, 1985).

9. Finally, residents expressed a belief that the Niagara Falls area had already done its share to bear the costs for regional waste disposal. Thus, equity was a major factor influencing the perceptions of local residents.

DISCUSSION

The opposing sides in the hearings offered diametrically different approaches to the task of projecting the impacts from the SCRF #6 facility. In this final section of the chapter, I will analyze the CECOS case study in light of the framework of limiting and expanding approaches to psychosocial impact assessment (see Table 9-2).

Definition of Psychological Impacts

In line with the limiting approach, the experts for the applicant confined their examination to the construct of stress, operationalized in terms of symptomology and studied in a controlled fashion. In contrast, stress was only one component of the interveners' expansive scope for psychosocial impact assessment. For the Intervener, the demands of projective impact assessment required that the expert form an opinion without the benefit of post-project data, and, in the case where changes have already occurred in the community, in the absence of detailed baseline information, as well.[5]

---

[5]The before-after comparison is simulated, for example, by using intensive focused interviews to reconstruct baseline residential quality prior to a toxic incident (Edelstein, 1982) and using this reconstructed baseline for comparison to the post-toxic incident circumstances. The target community can also be compared to similar places that have already undergone a change like that proposed, although the real degree of comparability is always subject to question in such comparisons.

It is noteworthy that the accuracy of impact assessment can be affected by change over time leading to different conditions than those available for study. Predictable changes include the evolution of the project through approval, construction, operation, closure, and long-term monitoring phases, with each phase

Table 9-2.
Expanding and Limiting Approaches to the SCRF #6 Case

Definition of Psychological Impacts

CECOS: Stress symptomology

INTERVENER: All future impacts on the mental well-being of people due to a project over time.

Definition of Causality

CECOS: Statistical causality

INTERVENER: Logical causality

Choice of Research Questions

CECOS: Is there a statistically greater degree of stress due to CECOS as compared to control, premeasure or normative data?

INTERVENER: What are the psychosocial impacts due to CECOS that are likely to continue should SCRF #6 be permitted?

Risk Assessment

CECOS: Assumes risks are minimal.

INTERVENER: Explores no incident and possible worst case scenarios over time.

Subjectivity Versus Objectivity

CECOS: Personalistic assumption.

INTERVENER: Environmental assumption.

Mitigation of Impacts

CECOS: Can be reduced through education and stress reduction techniques.

INTERVENER: Are inherent in the situation and cannot be mitigated due to historical loss of trust and local experience with hazardous waste facilities.

Statistical Versus Logical Causality

In line with the limiting approach, the applicant's experts argued that statistical causality is a requirement for valid research, established by

possibly offering its own idiosyncratic impacts. Less predictable is the variable of project performance, namely that SCRF #6 might perform as designed without incident, that it might fail in various minor ways (raising concern but not causing risk), that it might cause or contribute to various chronic and long-term problems and, finally, that it might cause or contribute to a catastrophic accident. Psychological impacts may also vary according to changes in contextual variables, such as the health of the local economy, that may change over time in unpredictable ways.

quantitative comparisons between pre-post measures, with norms for standardized measures or between the experimental and control groups. However, Covello noted:

> Even if scientists could agree on how to measure stress, it would be extremely difficult to demonstrate a direct causal relationship between SCRF 6 and significant levels of psychological stress in the community (Covello, 1988, p. 10).

In contrast, the intervener sought to **understand** the meaning of the facility to residents in light of their past experience and their concerns for future impacts, identifying logically consistent relationships between causes and effects. As a process observer, it was recognized that ultimately the expert, rather than statistically significant findings, must be the instrument of social impact research (Wolf, 1977).

The Research Questions

A given approach to research may be more appropriate for answering some questions and less appropriate for others. The **limiting approach** seeks to hold independent a few key variables while the **expanding approach** inductively searches for relationships. Thus, the CECOS experts raised the question: "Is psychopathology statistically greater today in Niagara Falls than in standard populations of test takers, in control communities or, historically, in Niagara Falls at an earlier point in time?" In contrast, the intervener asked: "What are the psychosocial impacts due to current CECOS operations that are likely to continue should SCRF #6 be permitted and what contextual factors are likely to influence these impacts?"

Risk Assessment

The CECOS argument rested on a presumption that risks from a state-of-the-art and secure facility (Covello, 1988, p. 29) are minimal and that, therefore, psychological impacts occur because perceived risk exceeds actual risk. In contrast, the intervener avoided assumptions about facility performance, arguing instead that the type and intensity of impacts would vary according to facility performance, but that even a successful facility would entail significant impacts.

Subjectivity Versus Objectivity

Much as in Metropolitan Edison v. PANE, the decision to hear testimony on psychological impacts from experts during the SCRF #6 hearings embodied the assumption that people cannot get past their own subjective states of mind and, therefore, experts are required to objectively evaluate whether psychological impacts have occurred. In this view, subjectivity is equated to self-interest, and the expert acts as a check against fraudulant claims. But how does the expert promote objectivity?

CECOS's experts maintained that the requisite for objectivity is the use of controlled research techniques. The interveners used their psychosocial expert to translate the subjective concerns of impacted people into objective terms that could be adjudicated.

Mitigation of Impacts

An apparent function for the Applicant's experts was to promote "appropriate" responses to hazardous technology by developing techniques for changing beliefs and "managing" stress or other impacts. Their approach was **personalistic** in that it focused upon changing affected people rather than changing the situation. Accordingly, CECOS's experts argued that stress related to SCRF #6 can be mitigated, either by individual stress management or through a series of "community-based risk communication and public education activities and programs" (Covello, 1988, p. 29), including tours and open houses and the circulation of brochures, video presentations, and a newsletter. Such activities were seen as helping residents to cope by increasing their understanding and familiarity with the facility. Over time, such activities, combined with community benefit projects, would provide "a firm basis for community trust and acceptance" (Covello, 1988, p. 33) that would reduce any psychological stress due to CECOS.[6]

---

[6]Covello relied on the successful siting of a hazardous waste facility at the Alberta Town of Swan Hill as an example of successful stress mitigation. However, there has been no evaluation of psychosocial impacts from that facility. Furthermore, major differences between the Swan Hill and SCRF #6 cases limit the validity of generalizations. Specifically, the Swan Hill facility was a new facility located in a low visibility area 10 km from a community lacking any history in dealing with hazardous wastes. It was a government project with a late and limited role for the private operator. The community controlled the siting process from the onset: the effort involved an invitational process driven by incentives and voluntary participation, and the final decision was dependent upon majority support in a plebicite in the host community (McQuaid-Cook and Simpson, 1986). In contrast, community perception of SCRF #6 is tainted by a long and negative local history with toxic sites, including the highly visible CECOS facility. Residents had unsuccessfully opposed previous CECOS permits and were strongly opposed to SCRF #6. Education was unlikely to overcome either the existing distrust of CECOS or the negative beliefs about the facility firmly established in the minds of a majority of nearby residents. The local population was all too familiar and knowledgable about the risks associated with hazardous technology.

In contrast, the intervener argued that psychosocial impacts such as those found near CECOS cannot be easily mitigated. This conclusion was based upon three factors. First, there is no evidence demonstrating that mitigation techniques work. Second, trust is the foundation for any mitigation program. Yet, the specific historical, social, and environmental context of Niagara Falls generally, and specifically with regard to CECOS, indicates that it is unlikely that trust in CECOS, in SCRF #6, or in the regulatory agencies can be easily established, conclusions well-supported by the Intervener's survey and interview data. Finally, educational efforts of the type suggested by Cavello may even backfire, enhancing rather than diminishing community concerns (see Slovic et al., 1982). In the interveners' **environmental** perspective, the project, not the people, is modified in light of potential adverse psychosocial impacts.

CONCLUSION

The CECOS case provides an opportunity to critically view the emergent field of psychosocial impacts of hazardous facilities. Despite a convergence of research findings evident in the literature (Edelstein, 1988), expert testimony diverges in hearings and trials along basic philisophical and adversarial lines. These disparities separate the issue according to cultures of psychology. The limiting and the expanding approaches mesh closely with these respective cultures.

Psychology as a Monoculture

CECOS's experts described psychology as a separable discipline that is consensually practiced as a behavioral monoculture focused on the construct "stress" and practiced through a controlled, quantitative, and positivistic approach used to identify the relationship between discrete variables. This quantitative paradigm is reductionistic, seeking to isolate variables for study. Assumptions are made about "control groups" based upon a few salient characteristics, but little is done to assure true comparability. Perhaps most importantly, the controlled approach is associated with a highly conservative definition of risk and causality governed by the equation of **significance** with **statistical significance**. As a result, actual causal relationships found in real field situations can easily be obscured.[7]

Psychosocial Impact Assessment as an Alternative View

The term "psychosocial," itself, reflects both the parallel evolution with psychology and sociology of cross-over subdisciplines, as well as the interdisciplinary focus inherent in impact assessment (NEPA, 1970). Psychosocial impacts should be studied using multimethod approaches that mix research techniques drawn from all the social sciences. The selection of the research approach depends on the situation, the context to be studied, the questions formulated, and the time, opportunities, and resources available for collecting information. Different contexts should be used to expand the scope of theory by attending to the differences, as well as the similarities, between superficially similar cases. It would be unfortunate to see so inherently diverse a field reduced to the quantified study of stress or, for that matter, to any other single hypothetical construct and set of standard measures.

Furthermore, one quickly learns that just because there is not a statistically demonstrated relationship between factors does not mean that they are not significantly related. Rather than allowing the term "significant" to be captured by the much more limiting connotation of "statistically" significant, the alternative of logical significance is suggested.

A further feature of the alternative expanding paradigm is its recognition and respect for the complexity of the real world (see Winkel, 1987). This recognition of complexity stands in stark contrast to the laboratory culture of controlled research, where problems are artificially reduced to a few variables. The challenge of psychosocial impact assessment is not to bring the precision of the laboratory into the field, an approach that does not meet the need of impact assessment. Instead, an expanding approach to the understanding of projects and their impacts is required.

---

[7]In science, one seeks to avoid errors where one concludes incorrectly that an effect was due to a cause. To avoid such "Type 1" errors, there is a tolerance of "Type 2" errors, where one disregards an effect due to insufficient proof (Edelstein, 1988; Levine, 1982). Statistical levels of significance are used to define the relative toleration of the two errors.

Implications

If the thesis of this chapter is correct, then philosophical conflicts within social science affect the definition of psychosocial impact assessment and the social application of social science to expert testimony. Within the adversarial system of decision making, the significance of such internal conflicts becomes confused with the advocacy of one side versus the other. As a result, the definition of psychosocial impact assessment relating to hazardous facilities is taken outside conventional professional discourse and placed before a "decider of truth." As legal precedent is set, the legal profession's selection of experts may well play a major role in defining how professionals, themselves, think of this field.

REFERENCES

1. American Psychological Association, Briefs of Amicus Curiae, U. S. Supreme Court, October Term, 1982.
2. Burdge, Rabel, Field, Donald, and Wells, Stephen. "Utilizing Social History to Identify Impacts of Resource Development on Isolated Communities: The Case of Skagway, Alaska." *Impact Assessment Bulletin,* 6 (2):37–54, 1988.
3. Concerned Citizens Organizations, Letter of Appeal, May 27, 1988.
4. Covello, Vincent. "Testimony of Dr. Vincent T. Covello on the Psychological Impacts of SCRF 6," prefiled testimony for Rivkin, Radler, Dunne & Bayh and Sive, Paget & Riesel, P.C. in the matter of the application of CECOS International, Inc. for the proposed Secure Chemical Residue Facility #6, application #90-85-0551, October, 1988.
5. Edelstein, Michael R. *Contaminated Communities: The Social and Psychological Impacts of Residential Toxic Exposure.* Boulder, Co: Westview Press, 1988.
6. Edelstein, Michael R. "Intervenor Concerned Citizens Organizations Prefiled Testimony of Psychological Impact Issues," report for Milbank, Tweed, Hadley and McCloy in the matter of the application of CECOS International, Inc. for the proposed Secure Chemical Residue Facility #6, application #90-85-0551, October, 1988 (b).
7. Edelstein, Michael R. "Disabling Communities: The Impact of Regulatory Proceedings." *Journal of Environmental Systems.* 16(2):87–110, 1986/87.
8. Edelstein, Michael R. "Toxic Exposure: Field Research and Expert Testimony." Paper presented at the Society for the Study of Social Problems, San Antonio, Texas, August 26, 1984.
9. Edelstein, Michael R. "The Social an Psychological Impacts of Groundwater Contamination in the Legler Section of Jackson, New Jersey." Report to the law firm Kriendler and Kreindler, 1982.
10. Edelstein, Michael R. "Social Impacts of Al Turi Landfill, Inc." Report prepared for the Town of Goshen, New York, 1980.

11. Edelstein, Michael R. "Participatory Futurism: Merging the Evolution of the Future with the Process of the Present," pp. 516–530 in Hnas Joachim Harloff (Ed.), *Bedingingen des Lebens in der Zukunft und die Folgen fur die Erziehung.* Berlin: Technische Universitat Berlin, 1978.
12. Edelstein, Michael R. and Wandersman, Abraham. "Community Dynamics in Coping with Toxic Exposure," pp. 69–112 in Irwin Altman and Abraham Wandersman (Eds.), *Neighborhood and Community Environments,* Volume 9 in the series Human Behavior and the Environment. N.Y.: Plenum Press, 1987.
13. Erikson, Kai. *Everything in its path.* New York: Simon and Schuster, 1976.
14. *Federal Register,* "Rules and Regulations." 49(236):47747, Thursday, December 6, 1984.
15. Gibbs, Margaret. "Psychological Dysfunction in the Legler Litigation Group." Report to the law firm of Kreindler and Kreindler, 1982.
16. Hartsough, Don M. "Legal Issues and Public Policy in the Psychology of Disasters," 283–307 in Gist, Richard and Lubin, Bernard (Eds.), *Psychological Aspects of Disaster.* New York: John Wiley and Sons, 1989.
17. Hartsough, Don M. and Savitsky, Jeffery C. "Three Mile Island: Psychology and Environmental Policy at a Crossroads." *American Psychologist,* 39(10): 1113–1122, 1984.
18. Hopkins, Michael. Letter to author of August 31, 1988.
19. Industrial Waste Siting Board, "Interim Decision," July, 1988.
20. Jordan, William S. III, "Psychological Harm after PANE: NEPA's Requirement to Consider Psychological Damage." *The Harvard Environmental Law Review,* 8(1):55–87, 1984.
21. Jorling, Thomas. "Interim Decision of the Commissioner," July 19, 1988.
22. Levine, Adeline G. *Love Canal: Science, Politics and People.* Boston: Lexington Books, 1982.
23. Marshall, Elliot. "NRC Must Weigh Psychic Costs: Environmental Law protects Mental Health, an Appeal Court Finds: Federal Attorneys See Broad Implications" *Science,* 216(11):1203–4, June 1982.
24. McQuaid-Cook, Jennifer and Simpson, Kenneth. "Siting a Fully Integrated Waste Management Facility in Alberta." *Hazardous Waste Management.* 36(9):1031–6, September 1986.
25. (The) National Environmental Policy Act of 1969, L. 91-190, 42 United States Congress 4321–4347.
26. New Jersey Supreme Court, A-83-84, decision in *Ayers v. Jackson Township,* May 7, 1987.
27. Pannell, Kerr, Förster. "Tourism Impact Study: Proposed OWMC Hazardous Waste Treatment Facilities." February 1985.
28. Pearlstein, Andrew, ALJ. Ruling of May 20, 1988.
29. Slovic, Paul, Fischhoff, Baruch, and Lichtenstein, Sarah. "Psychological Aspects of Risk Perception" pp. 11–42 in Sills, David, Wolf, C.P., and Shelanski, Vivian (Eds.), *Accident at Three Mile Island.* Boulder, Co: Westview Press, 1982.

30. Sorenson, John, Soderstrom, Jon, Copenhaver, Emily, Carnes, Sam, and Bolin, Robert. *Impacts of Hazardous Technology: The Psycho-Social Effects of Restarting TMI 1.* Albany: SUNY Press, 1987.
31. United States Supreme Court, Metropolitan Edison Co. et. al., v. People Against Nuclear Energy et. al., October Term, 1982,
32. *United States Law Week,* April 19, 1983, 51LW4371-5.
33. Winkel, Gary. "Implications of Environmental Context for Validity Assessments" pp. 71–98 in Stokols, Daniel and Altman, Irwin. *Handbook of Environmental Psychology,* Vol. 1. N.Y.: John Wiley & Sons, 1987.
34 Wolf, C. P. "Social Impact Assessment: The State of the Art Updated." *Social Impact Assessment,* 20, August, 1977.

Chapter 10

CHARACTERISTICS OF CITIZEN GROUPS WHICH EMERGE WITH RESPECT TO HAZARDOUS WASTE SITES[1]

E. L. QUARANTELLI

An increasingly noticeable feature of American society is the presence of newly formed groups of private citizens concerned with preventing and preparing for possible disasters or with recovering from actual impacts of such types of community crises. Their increased visibility and activity is probably reflective of broader trends in the country on the rights of consumers, an emphasis on participatory democracy, and an interest in organized self help, that are some of the legacies of the social turmoil of the late 1960s and early 1970s (for a partial examination of the historical background, see Boyte, 1980).

Apart from any changes in popular beliefs and values about the rights and obligations of individual citizens to work together, there has also been an increase in local community and formal advocacy groups interested in activating and mobilizing private citizens. Examples of the variety of such groupings are presented in Freeman (1983); they range from crime watch groups to those interested in neighborhood beautification. Thus, it is not surprising that there are also a variety of public interest groups across the nation who are consciously undertaking deliberate efforts to educate and train people in disaster preparedness, response, recovery, and mitigation. The larger changes in American society suggest that it should be anticipated that not only will emergent groups of citizens continue to surface in potential and actual disaster situations, but that their numbers are very likely to increase in the future.

---

[1]The work reported on in this chapter was partly supported by the National Science Foundation (NSF) under Grant #CEE-8113191. However, all opinions, findings, or conclusions expressed are those of the author and the Disaster Research Center, and do not necessarily reflect the views of NSF.

The development of new small groups has long been studied in sociology going back to almost the beginnings of the discipline (e.g., Coyle, 1930). Such work has continued into this decade (e.g., Ridgeway, 1983). Similarly, the emergence of new groups, especially after disasters, has long been noted (e.g., see descriptions in the first systematic sociological study of disasters, Prince, 1920). The coming into being of such groups has also more recently been the focus of direct study (see, e.g., Forrest, 1974; Bardo, 1978; Nigg, 1979) as well as theoretical examination (see, e.g., Kreps, 1984; 1985; Drabek, 1986). Also, there have been a few efforts to link the general sociological approach to new groups—usually within the subspecialization of collective behavior—and empirical disaster studies (see, e.g., Quarantelli, 1970; Stallings, 1978).

However, both strands of research have been very limited. Our general understanding of the development of new groups is not substantial, whether it be of new informal groups or of more formalized organizations. Likewise, although emergent behavior and groups have been looked at in the disaster area, the studies tend to be selective, and they also tend to focus more on the phenomena during the emergency time period of disasters rather than during the preimpact or postrecovery time periods. (For general discussions of the existing literature prior to 1980 see Quarantelli, 1985:2–10; Stallings and Quarantelli, 1986; the most recent analysis of the literature on emergent structures in disasters, written by Drabek [1987], became publically available after this manuscript was almost completed.)

Against this background, in late 1981, the Disaster Research Center (DRC) initiated an extensive sociological study of community-based citizen groups which emerged either to prepare for and/or recover from potential threats and actual disasters. The research extended over a four-year period and focused on local citizen groups who came into being outside of any immediate emergency period, and who were oriented to a full range of hazards from floods and hurricanes and tornadoes to nuclear and chemical plants and hazardous waste sites. In-depth interviewing was undertaken of members of over 50 such groups around the United States. Local organizational and community statistical and documentary data also were collected to supplement the interview data. A telephone follow-up survey of selected members was undertaken a year after the groups were first studied (for details on the methodology of the study see Quarantelli, 1985).

This chapter reports on the systematic DRC observations and findings on the dozen local citizen groups studied, which were organized with respect to hazardous waste sites. These were located in or around Blountville, Tennessee; Bumpass Cove, Tennessee; Deer Park, Texas; Knoxville, Tennessee; Milton, New Jersey; Oak Ridge, Tennessee; Pasadena, Texas; Toledo, Ohio; Wayne, New Jersey; Wilmington, Ohio; Woburn, Massachusetts; and Yellow Creek, Kentucky. More partial but relevant data were also obtained with respect to seven other citizen groups that had emerged in or around Cincinnati, Ohio; Cleveland, Ohio; Lake County, Illinois; Mentor, Kentucky; Milton, New Jersey; Montgomery County, Maryland; and Wayne, New Jersey.

Actually we found few significant differences among emergent citizen groups irrespective of the particular potential or actual disaster agent with which they were concerned. In fact, there was little difference in the orientation to either technological or nontechnological disaster agents, which some have speculated is an important distinction affecting the behavior of individuals and groups (see, e.g., Berren, Beigel, and Ghertner, 1980). But consistent with DRC studies on other topics, differences that exist are related to other than the physical source of the potential threat or risk such as the perceived potential dangerousness of the threat, its perceived uncontrollability, and a number of other factors.

In this research we studied the characteristics, the careers of and the conditions for emergent citizen groups (ECGS). But this paper primarily summarizes the conclusions reached about the first, namely, the characteristics of citizen groups that emerged with respect to hazardous wastes. The general research findings from the DRC two other foci of study have been reported elsewhere (see Green, 1983; 1984; Green, Neal, and Quarantelli, 1989; Neal, 1983; 1984; Neal and McCabe, 1984; Quarantelli, 1984; 1985; 1988).

We report on three major topics; the social compositions of ECGS that develop with respect to hazardous wastes, their structures, and their initial activities. In the last two cases, the summary of our empirical findings presented in propositional form, are subcategorized into more specific features. Under structure we indicate what was generally learned about different types of ECGs, and certain aspects of their division of labor, hierarchy, and formalization. Under activities are noted major general observations about the attempts to organized ECGs, their resource mobilization efforts, and how ECGs try to affect decisions and policies regarding hazardous wastes.

To avoid endless qualifications, propositional statements are presented in an *ideal type* format, meaning that all generalizations are advanced about the different phenomena as though the emergent citizen group (ECG) existed in pure form. Thus, while it is improbable that all of our observations would be found in everyone of the empirical cases we actually studied or in all ECGs, the depiction is as valid as can be made about an ideal type ECG or ECGs in general.

Similarly, we examined ECGS at different time periods of their development. Not all findings therefore are equally applicable for all time periods; generally we usually portray, unless otherwise specified, an ECG which has formed and has reached a point of formalization in structure, but which has not yet institutionalized; that is, which has not developed routine and established behavior. Our focus is on the emergent phase or formative stage of ECGs that have developed with respect to hazardous wastes.

Also, almost all propositions are advanced independently of most others even though there are often interactional effects among and between the behavioral aspects alluded to in each proposition. In another publication we attempt to relate and integrate a number of the more important factors into a systematic model of the phenomena (see Quarantelli, 1985:48–51).

Finally, it is necessary to note that DRC gathers all its data with the understanding that individuals who participate in the field work will never be identified in any personal or identifiable way in any public report of the Center. This promise of confidentiality therefore precludes a reference to any particular emergent citizen group, for given their size, it would be easy to identify particular members. This DRC policy may take away from the vividness of any account of the groups, but it assures surprising candor and honesty by persons who are interviewed. In addition, our emphasis on the general and the generic is reflective of a sociological rather than a specifically historical interest about emergent citizen groups, and of a concern with such groups as may emerge in the future instead of those that came into being in the past.

COMPOSITION

The typical ECG has less than a hundred members, but the range is from a dozen to several thousand people. Since formal membership rosters are seldom kept by the groups, the number of members is almost always an estimate and probably an overestimation in terms of persons who consider themselves members. Participation in ECG activities does not correspond to what might be indicated by formal (e.g., dues paying) or psychological (i.e., identification with the group) membership. In general, there are more participants, over time, than members.

There are usually three general kinds of participants in ECGS: a small but very active core; a somewhat larger supporting circle who can be mobilized for specific tasks; and, a great number of primarily nominal supporters who may pay dues, receive newsletters, and attend an occasional meeting. In some cases, nominal supporters do not even consider themselves group members, but may nonetheless participate, e.g., by signing a petition to remove a hazardous waste site from a particular locality. Many ECGS have nonmember participants such as public officials, technical professionals, or mass media reporters who provide information, knowledge, advice, or other resources to the group because of their sympathy with the group, but who are not formal members of the group.

In all ECGs there is a very active core of members; seldom more than a half dozen in number. In some cases the core is all there is insofar as active membership is concerned. Outside of that core, participation in group activities tends to be episodic and sporadic in most groups. In fact, for lower level participants, who are under practically no group surveillance, there are only very diffuse role expectations.

Participation in ECGs is almost always a very part-time activity. But core members often devote extremely large blocks of time to the group, and the work may be more than full-time for an occasional core member or two. In some rare instances, the whole life of the core member is oriented around ECG activities.

Core members are usually early joiners of ECGs, and tend to remain in the group for very long periods of time. There is little turnover of core members except for some occasional "burn out" cases, or as a result of moving out of the area (for reasons independent of the hazardous waste problem focus of the ECG). Core members tend to have only very general perceptions of most other members; whereas core members are very salient to other members who often have only general impressions of others in the group. Outsiders who deal with ECGs, such as community officials or mass media personnel, seldom distinguish between perceived membership and participation, or recognize different kinds of participants in the ECG.

In most cases ECGs have a disproportionate number of women members. The core and its leadership is also disproportionately female in the great majority of the groups. While both partners in married couples are often members of the same ECG, one partner, frequently the male, is usually less active.

While all adult age ranges are represented in ECGs, the typical member is in the 30 to 40 age range. Perhaps contrary to some popular conceptions, retirees are not prominent as a whole in ECGS, although an occasional retiree may have a key role in an ECG usually because of possession of specialized knowledge regarding hazardous wastes and which are relevant to the group activities.

ECGs are drawn primarily from the white collar middle class. But they sometime involve mobile members of the working class: that is, individuals from blue collar segments. The lowest socioeconomic levels are seldom involved in emergent groups and never comprise the core or more active membership. There are occasional ECGs made up of upper middle and lower class background. These tend to be more structurally complex groups than those with other social class composition. Minorities in American society are very poorly represented in ECGs although they often reside near or around hazardous waste sites.

Although not all ECGs are neighborhood based, many draw their members from a particular neighborhood. This usually occurs where the hazardous waste issue or problem around which the ECG is focused is neighborhood specific. ECGs involved in more community-wide issues draw from a wider geographic base, but there is still a tendency for membership to be drawn from clusters of specific neighborhoods.

Overall, membership in ECGs tend to reflect lifestyle (i.e., social class position, social linkages, and social experiences) more than it does personality or demographic characteristics. This seems to be true of all levels of membership.

STRUCTURE

There are at least two major types of ECGs. The first are the specific task-oriented groups which are likely to be but not exclusively *post*disaster groups (e.g., after a hazardous waste incident has resulted in problems requiring some kind of response), and are focused primarily on personal and self-interests of their members (e.g., loss of property values of their homes). The second are broader community-oriented groups which are more likely to be *pre*disaster groups, and are concerned mostly with raising community awareness of a possible threat or disaster from an hazardous waste site. The first type has limited goals and is inclined to have exclusive membership from a particular neighborhood. The second type is more likely to have open-ended goals, and tends to have inclusive membership (e.g., anyone in the involved community, whether potentially threatened or not by the hazardous waste threat, can join).

The specific task-oriented ECGs tend to be centralized in a neighborhood or area; the more community-oriented groups tend to draw their members from the community generally although not from all *areas* of a community (because of the social class composition of most ECGS). Task-oriented groups, on the average, are smaller than community-oriented groups.

Another major distinction between ECGs is between those engaged in conflict with other groups, and those in nonconflict situations. The great majority of ECGs exist in conflict situations. A cross-classification of the conflict/nonconflict dimensions and the orientation dimension results in a fourfold typology of ECGs as follows:

(1) nonconflict task-oriented ECGs,
(2) conflict task-oriented ECGs,
(3) nonconflict community-oriented ECGs, and
(4) conflict community-oriented ECGs.

Division of Labor in ECGs

The core of emergent groups, whatever the type, almost always has a division of labor, often in terms of the particular personal skills core members possess. The division of labor, therefore, is often sharp because roles are not easily interchangeable. Also, the division of labor tends to be more elaborate in community-oriented ECGs. The division of labor in ECGs has more to do with externally-oriented behavior than with internally-oriented behavior. This reflects the strong instrumental activities of ECGs regarding hazardous wastes and their weak group maintenance activities. But there is seldom any division of labor beyond the active core and its supporting circle in the largest of ECGs. The division of labor elaborates only up to a certain point in most ECGs; in some cases, a more complex division of labor may actually be replaced by a simpler one contrary to certain general principles of organizational development (see, for example, the discussion of the iron law of oligarchy presented in Michels, 1915).

Hierarchies in ECGs

There seldom is any actual hierarchy in the group core even though there may be a formal hierarchial order as a result of having formal officers or positions. The exception to this is when there is a charismatic leader who often is the original founder of the group. In that case, such leaders have more influence than others in the core. Degree of influence also appears to be related to the ability of core members to mobilize certain kinds of resources.

The formal or official hierarchy of ECGs does not necessarily reflect core membership or different degrees of influence in the core. The formal officers are often not the informal leaders and infrequently have no public visibility.

Leadership often is given to rather than taken over by initial core participants: that is, it evolves slowly and informally. Leadership, moreover, is fairly stable in most ECGs especially among those who are informal leaders or members of the core. There is both internal group pressure and self-imposed pressure to "downplay" leadership. The word "leader" is often avoided and emphasis is placed on the supposedly democratic nature of the ECG.

Conflict groups tend to somewhat less democratic in procedures and are more hierarchial in structure than non-conflict groups. Furthermore, conflict groups are more vertically and horizontally structured than nonconflict ECGs.

Formalization of ECGs

ECGS with higher-level socioeconomic members start out more organized and formalized than groups with primarily lower level socioeconomic members. There are varying degrees of formalization. Some ECGs only develop an informal structure; a greater proportion set up a formal organization, and most formally incorporate if they have any life career at all.

Formal incorporation means that the ECGs have a charter, formal group positions, and nominally at least an initial membership roster. Incorporation also generates a certain amount of bookkeeping, leads to the opening of a bank account, and the use of letterhead paper.

Formalization does not seem to be related to task or community orientation, but conflict groups tend to be more formalized. Outsiders tend to take seeming symbols of formalization such as a group name or a title used by a member (e.g., chair of a committee) as an indication of ECG formalization, whether this is the case or not.

ACTIVITIES

Among the major activities of ECGs are attempts to organize the group, to mobilize resources, and to bring about decisions and policies favorable to the group. ECGs seem to seldom engage in primarily symbolic or expressive actions; they are heavily instrumentally oriented. The possible exception to this are some of the antinuclear plant groups which can draw from an outside ideological base which often stresses symbolic resistance to nuclear power.

The major activities of ECGs are carried out by the active core, but in the majority of ECGs the core can regularly mobilize a significant proportion of the nonactive members for a public display of numbers. Examples would be getting the usually nonactive members to appear for a special meeting of the group, to participate in some public activity of the ECG, and to write letters.

ECGs have far more internal disagreements and conflicts about what courses of action to follow than are usually publically visible. Internal differences are played down. Strong dissenters tend to leave the ECG rather than to create a schism and a new group.

Some ECGs are peopled by newer residents in an area. This sometimes leads to a clash or confrontation with longer time residents seen as controlling and/or not effectively using the local governmental structure to solve the perceived problem regarding hazardous wastes. Especially in smaller communities this may lead to a wider community conflict between the newcomers organized in an ECG, and longer established residents. In such cases, activities broaden out to those relevant to community cleavages, and go beyond those involved in a dispute over a controversial problem.

Organizing the Group

Organizing ECGs involves an early clarification of goals and objectives, and the development of initial strategies and tactics. Although such organization is a never-ending activity, it is not visualized by the first core members. In fact, it usually takes a relatively long time before there is a recognition that goals and objectives, and strategies and tactics may have to be modified or changed often during the career of the ECG.

Goals

Most ECGs initially have only very broad and vague goals such as "being able to live in a safe place." Such goals, since they involve matters of security and health, which usually directly affect the family home and life of ECG members, are implicitly deemed unassailable or unchallengeable by anyone, and certainly not by public officials or agencies with implicit if not explicit responsibilities for the safety of citizens. Thus, ECGs view a typical initial goal as locating the responsible authorities who can take action necessary to solve the hazardous waste problem.

Almost all issues raised by ECGs in initial approaches to outsiders are perceived as being ignored or rebuffed, or as resulting in reactions not addressing their issues. This is often a correct perception. In conflict situations in particular, ECGs are initially and frequently seen by those private or public sector officials approached, as being uninformed or narrowly biased about the hazardous waste issues, and unrealistic or simplistic in solutions proposed or goals sought. This too is often a correct perception.

Redefinition of goals frequently occur after early group emergence. Goals of ECGs are far more likely to expand or change than to contract or remain static.

A major manifest activity of almost all ECGs once formed is awareness creation. Although perhaps not originally, the awareness creating function comes to be perceived as very important, and will be maintained even if other goals are changed. Core members come to believe strongly that creating or maintaining awareness of the threat or danger they perceive from hazardous wastes is crucial to holding the interest of those already members of the ECG, to obtaining more recruits for the ECG, and to convincing those officials who need to be informed and impressed about the problem around which the ECG is focused.

Initial Strategies and Tactics

Redefinition of the goals of ECGs also often involves a reconsideration of the means that the ECG should use. In some cases, the focus on means may overshadow the old or new goals, as matters of strategies and tactics come to the fore.

Questions rather than demands constitute the bulk of the initial communications from ECGs to governmental agencies; demands only appear later. ECGs often have little idea where decisions relevant to hazardous waste problems are made, and thus many early actions are often misdirected. In time, some core members usually get fairly knowledgeable about the organizational decision-making process in their communities, but correct identification of sources of power does not necessarily translate into the evoking of desired decisions or policies.

While intended results do not always follow, many members of ECGs appear to believe that being a "squeaky wheel" is an appropriate strategy, as long as the actions undertaken will not be interpreted as radical by the larger community. While usually avoiding confrontation, most ECGs appear to prefer operating in public rather than working behind the scenes.

Many ECGs undertake a great deal of correspondence, especially initially; phone calls and personal visits to officials tend to occur later. Conflict ECGs in particular and especially core members sometime learn to use mass media reports about hazardous waste problems as pressure on officials; in some cases such news stories have been provided by mass media personnel who are covert supporters of the ECGs involved.

Common sense attitudes and beliefs about the value and helpfulness of different strategies seem to dictate what will be used, far more than productivity of earlier usages or availability of resources.

Mobilizing Resources

Resource mobilization includes recruiting new members, holding meetings, distributing newsletters, and obtaining resources. The last, however, is important in a way different than it might be usually visualized.

Recruitment

Recruitment of new members is seldom given a high priority in the great majority of ECGs. Such recruitment, as is undertaken, tends to be sporadic, haphazardous, and unorganized. Recruitment into some ECGs is handicapped because some potential members see public emphasis on the group problem as possibly generating more personal problems, such as reducing real estate values, for example, by emphasizing the nearness of a toxic waste site, which might make it more difficult for them to relocate later if the collective problem is not solved.

There is a tendency for early ECG recruitment to be for more members, and for later group recruitment to be for specific expertise. But the recruitment potential problem differs somewhat in the two types of

ECGs mentioned earlier. Task-oriented groups usually have a delimited number of people they could recruit, but in most cases they do not have to convince people there is a problem (e.g., the existence of an hazardous waste site in the neighborhood). Community-oriented groups typically have a much wider number of people they could potential recruit; however, frequently they have to convince potential recruits there is a problem for them as, for example, when the hazardous waste site is not in their neighborhood.

Meetings

In most ECGs, meetings are regularly held at least by the core; larger membership meetings are held far less often. There are also far more informal than formal meetings, and it is at the former that decisions are usually made and policies are typically set.

Core group decision making is almost always informal and highly democratic except in such instances where there is a charismatic leader. In some cases, agendas for meetings are usually preset by the core although nominal democratic procedures extend to all group activities. More often the core uses the larger ECG meeting primarily to ratify core decisions.

Formal ECG meetings are usually informally run with little attention to parliamentary procedures. Voting by balloting is rare, and secret voting almost nonexistent. Decisions by seeming consensus is the norm. While concern over obtaining larger group approval is genuine among most cores, lack of overt objections at meetings is often taken as a sign of approval of what actions have been proposed.

Formal meetings tend to be held more often during the early stages of group development. Turnout of meetings drops off substantially after a while, but some especially relevant happening may generate a high turnout for a particular meeting.

Newsletters

Most ECGs attempt to provide a newsletter or some publications for their membership. The production of such material is usually the creation of only one or two persons. Newsletters are used as a device for the dissemination of information desired by key core members. The longer an ECG exists, and the higher the socioeconomic backgrounds of members, the more likely a newsletter is published.

Resources

The great majority of ECGs have very little money, but they also need very little to operate. Funding is not a major problem for the typical ECG, because even though money could be used, it is normally not crucial. Most ECGs generate funds primarily from dues and voluntary contributions of members; this is sometimes supplemented by money obtained from informal activities such as bake or garage sales or car washes.

Money is far less important as a group resource than are nonmaterial factors such as information, specialized knowledge about hazardous waste, access to key persons, being able to meet, etc. Meeting space for most groups is sometimes provided by established religious groups who otherwise are seldom important in the development of ECGs. Nonmonetary material resources, such as paper for newsletters and typing assistance, are primarily obtained through the voluntary donations or offers from or through members of the ECG.

Sympathetic local college or university faculty members sometimes are sources of specialized knowledge, especially about the specific nature of the hazardous waste threat with which the ECG is concerned. In addition, one or two core members will often as a result of individual reading, library or newspaper research, and/or finding of knowledgeable individuals, become a considerable repository of relevant knowledge for the group.

Extremely few ECGs are able to obtain grants from either public or private sources. Occasionally they directly or indirectly gain access to community development funds, but almost always their operations are outside of the criteria necessary for most grants.

Affecting Decisions and Policies

Affecting decisions and policies includes identifying the officials and organization who might be able to do something about the perceived problem of hazardous waste, actually contacting the relevant parties, and joining in like or common efforts with other groups.

Identifying Relevant Groups

The initial general approach of ECGS to other groups seems to rest on the assumption that some definite group or official out there ought to be able to help the group, to use words our respondents used. Almost without exception, ECGs initially have little knowledge about whom they should approach with their problems. Also, the help sought in the initial shotgun approach is often undefined and unclear to the ECGs themselves.

Initially ECGs use a shotgun approach in approaching groups, organizations, and agencies which they think might be able to help them in some way. The consequence is that a variety of public and private groups are approached, as well as officials at different levels (usually those more visible). An occasional knowledgeable core member can short circuit the shotgun approach, but even sophisticated individuals often find it difficult to identify who should be approached first regarding hazardous waste issues.

Contacting Others

Initiative in making contacts with other groups and organizations is usually taken by the ECG. At times, after the group is informed, it may be contacted by some hazardous waste-oriented national level organization or public interest group. Governmental agencies and officials almost always wait to be contacted by ECGs.

There are major quantitative and qualitative differences in ECGs in conflict as compared with nonconflict situations. The former tend to attempt a greater number of, and usually more powerful organizational contacts than the latter. Most ECGs, especially those in a conflict situation, make repeated contacts with organizations they think might be relevant to hazardous waste issues. Initial lack of a response or an inadequate response does not serve as much of a deterrent to later contacts. There is a strong tendency for many ECGs to keep previously approached groups and individuals on a mailing list, and to approach them in a letter-writing campaign or petition submission, even when earlier approaches have not been fruitful.

Elected, more than appointed, officials are more likely to go through the motions of listening to questions and complaints from ECGs. At least they are more likely to provide some kind of feedback, such as an acknowledging letter.

When women are core leaders there often is an internal core perception that the ECG is at a disadvantage in dealing with bureaucracies and governmental units. In some cases, this is a correct perception because some male as well as female officials do tend to discount women leaders.

Private organizations which become the object of attention of ECGs do not appear to differ substantively in their reactions from public or governmental organizations. But they sometimes mount a seemingly more systematic public relations campaign in response to the issues raised by hazardous wastes. In some localities and in certain sections of the country, the private organizations involved in conflicts with ECGs often have the little disguised support of some governmental entities at the local and/or state level.

Joining With Other Groups

Local ECGs almost always avoid identification with the traditional established political parties in the community. This partly reflects the differing political affiliations or leanings of the membership of typical ECGs. One consequence is the avoidance of cross pressure on many ECG members.

The hazardous waste-oriented ECGs generally maintain a single issue posture, leading to their reluctance to align with other local and extra local groups with different goals, since that might lead to diffusion of group attention or surfacing of differences of opinion on other controversial issues.

While a single issue ECG reflects the typical situation, there are instances of multiple ECGs oriented toward similar hazardous waste problems within a given community. In multiple ECGs situations, coalitions may be formed among the ECGs involved. However, such groups in the same locality are more likely to cooperate than to develop coalitions, even if involved with the same problem. ECGs usually demonstrate strong internal concern over losing their autonomy of action.

In a multiple ECGs situation, core members of the different ECGs will sometime join together to form an umbrella type community-wide organization concerned about hazardous wastes. This frequently results in loss of public visibility of the local or neighborhood ECGs as outsiders tend to respond to the larger umbrella organization. While some umbrella organizations form for the purpose of disseminating information, others attempt to bring about common, often explicit, political action. In the latter case, the umbrella organization may become part of or be associated with established community action groups. This is not always functional for the local hazardous waste ECGs involved as their particular concerns becomes subordinated to larger but often distant issues.

More typical is for local ECGs to develop extensive horizontal networking through contacts with and at times establishment of coalitions *across* rather than within communities. Horizontal networking with other local emergent groups is important. Ideas of how to proceed and who to contact regarding hazardous waste issues and problems are often derived from such networking. Specific core members are often appointed/designated as boundary personnel with these other organizations.

In coalition or cooperative situations, credit is sometimes claimed by an individual ECG for what has been done collectively. Whether in a coalition or cooperative effort, the contact between the organization is usually undertaken by a few core members of the participating ECGs. Mergers of ECGs is almost unknown. Even at public hearings on hazardous waste issues, where many groups are represented, members of different ECGs tend to sit apart with members of their own group.

SOME POLICY IMPLICATIONS

There are at least three policy implications in the observations and findings presented. First, citizens feel that their complaints about perceived hazardous waste problems should be taken seriously by community officials. Initially people are more than willing to help such officials deal with the problem. They only become hostile and angry when their complaints are ignored or downplayed. Officials can easily avoid a conflict relationship by early and reasonable attention to the problems called to their attention.

Second, those persons interested in getting something done about neighborhood and community problems at least in the hazardous waste area, need to recognize that the leadership and necessary actions will have to be undertaken by only a handful of individuals. It is possible to catch the attention of many and to mobilize fellow citizens for occasional and certain activities, but the bulk of the work will have to be done by a few. Just as revolutions are the handiwork of a relative few, so are successful campaigns to bring about something important regarding hazardous waste in a neighborhood or community. To wait around for the "masses to rise in revolt" is to ensure that nothing will happen.

Third, the more the potential and actual members of ECGs, as well as the organizational and community officials who have to deal with them, are knowledgeable about the process that is involved as described in this chapter, the better off everyone will be when the situation starts to develop. Neither ECGs nor officials can be expected to have such knowledge on the basis of their own limited experience, so they cannot be very helpful in that respect. Researchers, while they may have the knowledge, are not usually good and often not very interested at showing the applications of their findings. However, other groups, whether public or private, can be very useful by calling attention to what good studies and research have found. Put another way, there is an important role they can play regarding the transfer of social technology which cannot be done by researchers nor by those citizens and officials who need the knowledge.

A CONCLUDING OBSERVATION

We have depicted the general characteristics of emergent citizen groups oriented to hazardous wastes. While some of the findings were unexceptional, a number certainly were unanticipated and/or contrary to popular conceptions if not prevailing social science hypotheses. As such, we have taken a step forward in our understanding of the phenomenon of ECGs. However, it is just that, a first step.

The research results will have to be tested with more and different groups. In particular, it will be necessary to see how the observations hold up in different societies. Only when systematic knowledge about the universal and the societally specific characteristics of ECGs is available will we be able to say that our knowledge rests on solid ground that will be useful both for practical and theoretical purposes.

REFERENCES

1. Bardo, John: Organizational response to disaster: Typology of adjustment and change. *Mass Emergencies, 3:*87–104, 1978.
2. Berren, Michael R., Beigel, Allan, and Ghertner, Stuart: A typology for the classification of disasters. *Community Mental Health Journal, 16:*103–120, 1980.
3. Boyte, Harry: *The Backyard Revolution: Understanding the New Citizen Movement.* Philadelphia, Temple, 1980.
4. Coyle, Grace: *Social Process in Organized Groups.* New York, Smith, 1930.
5. Drabek, Thomas: *Human System Responses to Disaster: An Inventory of Sociological Findings.* New York, Springer Verlag, 1986.
6. Drabek, Thomas: Emergent Structures. In Dynes, R. R., De Marchi, B., and Pelanda, C. (Eds.): *Sociology of Disasters: Contributions of Sociology to Disaster Research.* Milan, Franco Angeli, 1987, pp. 259–290.
7. Forrest, Thomas: *Structural Differentiation in Emergent Groups.* Newark, DE., Disaster Research Center, University of Delaware, 1974.

8. Freeman, Jo (Ed.): *Social Movements of the Sixties and Seventies.* New York, Longmans, 1983.
9. Green, Kenneth: A case study analysis of the relationship of local newspapers and disaster related citizen groups. Preliminary Paper # 86. Newark, DE., Disaster Research Center, University of Delaware, 1983.
10. Green, Kenneth: Implications of rural-urban differentiation: A study of local grass roots organizations in disaster situations. Ph.D. Dissertation. Columbus, Ohio, Department of Agricultural Economics and Rural Sociology, Ohio State University, 1984.
11. Green, Kenneth, Neal, David, and Quarantelli, E.L.: An examination of the relationship of emergent citizen groups to other organizations. Preliminary Paper #134. Newark, DE., Disaster, Research Center, University of Delaware, 1989.
12. Kreps, Gary: Sociological inquiry and disaster research. *Annual Review of Sociology, 10:*309–330, 1984.
13. Kreps, Gary: Disasters and the social order. *Sociological Theory, 3:*49–65, 1985.
14. Michels, Robert: *Political Parties: A Sociological Study of the Oligarchical Tendencies in Modern Democracy.* New York, Hearst, 1915.
15. Neal, David M: Types and functions of umbrella organizations for local social movement organizations: A look at emergent citizen groups in disasters. In Larkin, M., Honnold, J. and Williams, J.S. (Eds.): *Sociological Research Symposium XIII.* Richmond, VA.: Department of Sociology, Virginia Commonwealth University, 1983, pp. 119–122.
16. Neal, David M.: Blame assignment in a diffuse disaster situation: A case example of the role of an emergent citizen group. *International Journal of Mass Emergencies and Disasters 2:*251–266, 1984.
17. Neal, David M. and McCabe, Susan: Emergent citizen groups in disasters and their political activity: A look at natural hazard situations. Preliminary Paper # 90. Newark, DE., Disaster Research Center, University of Delaware, 1984.
18. Nigg, Joanne: The emergence of issues and collectivities: Community response to earthquake prediction and its consequences. Ph.D. Dissertation. Los Angeles, CA., Department of Sociology University of California, 1979.
19. Prince, Samuel: *Catastrophe and Social Change.* New York, Columbia University Press, 1920.
20. Quarantelli, E. L.: Emergent accommodation groups: Beyond current collective behavior typologies. In Shibutani, Tamotsu (Ed.): *Human Nature and Collective Behavior.* Englewood Cliffs, NJ: Prentice-Hall, 1970, pp. 111–123.
21. Quarantelli, E. L.: The preparation of citizen groups for earthquakes: The atypical nature of such groups and the conditions for their emergence. In *Proceedings of the Eight World Conference on Earthquake Engineering, Volume 7.* Englewood Cliffs, NJ: Prentice-Hall, 1984, pp. 901–908.
22. Quarantelli, E. L.: *Emergent Citizen Groups in Disaster Preparedness and Recovery Activities.* Final Project Report #33. Newark, DE., Disaster Research Center, University of Delaware, 1985.

23. Quarantelli, E. L.: Citizen groups and hazardous waste: Their careers and conditions for emergence. In *Hazardous Waste Detection, Control, Treatment: Proceedings of the World Conference on Hazardous Waste.* Amsterdam, The Netherlands, Elsevier, 1988, pp. 1757–1764.
24. Ridgeway, Cecilia L.: *The Dynamics of Small Groups.* New York, St. Martin's Press, 1983.
25. Stallings, Robert: The structural patterns of four types of organizations in disaster. In Quarantelli, E. L. (Ed.): *Disasters: Theory and Research.* Beverly Hills: CA., Sage, 1978, pp. 87–103.
26. Stallings, Robert and Quarantelli, E.L.: Emergent citizen groups and emergency management. *Public Administration Review 45:*93–100, 1985.

Chapter 11

PUBLIC ACCEPTABILITY OF ALTERNATIVE HAZARDOUS WASTE MANAGEMENT SERVICES[1]

RAE ZIMMERMAN

INTRODUCTION

The scale and distribution of service facilities have been debated for almost every large publicly-regulated service, including the handling of wastes, and these debates often have deep social and psychological overtones. In the decade of the 1980s, this issue has emphasized whether to decentralize or centralize services in many diversified areas of managing wastes or residuals. The same arguments arise for the storage, transport, treatment, and disposal of different kinds of residuals—solid wastes, sewage sludge, medical wastes, and hazardous wastes—since they often compete for the same kinds of solutions. This issue of the centralization or decentralization of waste handling has been a management problem ever since the size and density of population settlements exceeded the natural capacity of the environment to assimilate wastes at the point at which they were generated.

The implicit and explicit choices the public makes regarding the scale and distribution of facilities in one major service area—hazardous waste treatment and disposal—are examined here from a social-psychological perspective. The social-psychological perspective taken here supplements the traditional technical and economic approaches to this issue. It will be argued that onsite (decentralized) hazardous waste treatment and disposal facilities tailored to the wastes of individual establishments have become implicitly and explicitly acceptable to the public. This argument will be based, first of all, upon the fact that such facilities are very prevalent throughout the country, in spite of public policies and market forces aimed at centralization. Second, onsite facilities that provide permanent waste treatment are receiving implicit support from a wide range of interest groups. Third, the level of public controversy, as indicated by the intensity of media coverage and other factors, for the onsite option is low, once the treatment technology for the site is chosen. If public acceptance, in fact, exists for this particular option, it has considerable implications for the way alternatives are proposed and regulated in the future, given the role that public acceptability has in decision making for environmental facilities and the number of facilities that will have to choose between these two options. Another purpose of this chapter is to explore the conditions under which acceptance of relatively smaller scale or decentralized, onsite facilities actually occurs. In fact, many of these conditions may reflect important public objectives in their own right, and size and distribution of hazardous waste facilities may only be intermediaries or a means to achieve those objectives rather than desirable ends in themselves.

In order to explore this issue, the overall magnitude of the hazardous waste problem in the United States is, first, briefly described along with what decentralization and centralization mean in the hazardous waste management service industry and how prevalent decentralization and centralization is in that industry. Second, the positions that are emerging at all levels of government and among interest groups with regard to the two options are discussed. Third, the social-psychological approach is conceptualized, along with how this approach differs from the traditional technical and economic approaches. As a way of further exploring the advantages and disadvantages of the two options in social-psychological terms, concepts that characterize the social-psychological perspective are operationalized and applied in general terms to hazardous waste facilities regulated under the Resource Conservation and Recovery Act (RCRA) and its Hazardous and Solid Waste Amendments of 1984 (HSWA) and the Comprehensive Emergency Response Compensation and Liability Act (CERCLA) and its Superfund Amendments and Reauthorization Act (SARA) of 1986. These two laws define two major federal programs that shape hazardous waste management in both federal and state government in the United States. Fourth, results of an empirical analysis are presented of the relative degree of controversy at decentralized and centralized RCRA and CERCLA sites. Examples are presented that highlight some of the circumstances under which controversy arises over onsite facilities. Finally, the conditions under which decentralized facilities are publicly acceptable as a hazardous waste management option are identified and explained.

Magnitude and Nature of the Hazardous Waste Problem

Hazardous waste management is in a state of crisis. The industry is faced with legislative deadlines, increasingly stringent constraints on the technologies allowed,[2] and a growing stockpile of material for treatment and disposal. While estimates are hard to obtain that consistently define and inventory hazardous waste, some order of magnitude consensus is appearing in the literature. An estimated 250 to 280 million tons of hazardous wastes are generated each year that are regulated under RCRA/HSWA.[3] The National Academy of Sciences[4] gives a figure of 264 million metric tons. A later estimate places the 1985 figure at 247 million metric tons of RCRA regulated wastes with an additional 322 million metric tons of hazardous wastes not regulated under RCRA.[5] With new disposal and treatment requirements, the estimated percentage increase in wastes by 1990 is put at about 15 percent (not including the effects of recycling).[6]

When hazardous waste cleanup activity associated with abandoned sites regulated under CERCLA/SARA is considered, the problem is even greater. The inventory of hazardous waste sites that are potential sites for remediation grew from an initial 13,000 sites in 1982 to over 26,000 sites by 1986.[7] It has been estimated that by 1986, the U.S. EPA's CERCLIS inventory of potential sites for regulation under Superfund had been growing at the rate of 2,000 to 3,000 sites a year, and the number of sites being proposed for cleanup under the National Priorities List (NPL) was about 400 sites per year through 1985.[8] As of the first quarter of 1989, 1,165 sites from that inventory were targeted for cleanup under the National Priorities List (NPL), covering close to a nine-year history of the Superfund program.

Centralization and Decentralization of Services in the Context of Hazardous Waste Management

Most hazardous waste management facilities in the United States are regulated either as operating facilities under federal or state programs in conformance with RCRA/HSWA or as facilities that are abandoned or otherwise neglected, and are in need of cleanup under CERCLA/SARA. The distinction made here between onsite facilities (also referred to as individualized or decentralized facilities) and offsite (or centralized) systems depends upon the program under which the facility is being regulated.

RCRA/HSWA Regulated Facilities

Facilities regulated under RCRA/HSWA are operating facilities or facilities seeking permits to operate hazardous waste treatment, storage, or disposal facilities. For these facilities, the distinction between decentralization or centralization is made on the basis of whether or not wastes from different geographic areas are being transported to a single site for treatment and/or disposal.

Onsite or decentralized management refers to waste treatment at a single source, such as a factory, for wastes originating from that source involving no transport of raw wastes except within the property boundaries of that source. They are considered decentralized, since wastes are processed individually at each source rather than at another location for treatment and disposal. These onsite or decentralized systems may involve the temporary or permanent importation of waste treatment technologies to the site of generation.

Centralized facilities, in contrast, process hazardous wastes that originate from locations other than the site of the treatment or disposal facility, requiring those raw wastes to be transported to the site. These facilities are typically commercial operations.

CERCLA/SARA Regulated Facilities

Onsite facilities regulated under CERCLA/SARA (also called Superfund), refer here to those sites where remedial, long-term, and permanent cleanup efforts are occurring on the site, and waste material is not being removed to an offsite treatment facility. While many interim measures are being used that involve temporary onsite storage or soil and groundwater treatment, these are not considered among the CERCLA/SARA onsite systems in this chapter unless they become permanent remedies.

Offsite treatment refers to those sites being cleaned up through the removal of wastes to an offsite facility, which ultimately treats and disposes of the wastes. U.S. Environmental Protection Agency (EPA) policy requires that the offsite treatment of Superfund site wastes occur at facilities that are approved under the RCRA permit program.

The Social-Psychological Perspective

Over time many different perspectives and viewpoints have been used to evaluate the relative advantages of onsite (individual or decentralized) vs. offsite (centralized or commercial) alternatives for waste management. The relative attractiveness of the two alternative waste treatment systems was often argued in terms of technical or deterministic models that focused primarily upon engineering efficiency. This tradition of analysis has existed since such engineered systems were marketed. This approach was radically altered by the requirements of the National Environmental Policy Act (NEPA), which broadened the evaluation criteria for such systems to include environmental and social impacts as a basis for formulating and choosing alternatives. Later, the use of quantitative risk assessment followed, which broadened evaluation criteria still further, primarily in the direction of health effects. Concurrently, arguments of environmental economists were presented in terms of economies of scale, which usually ended up in support of centralization. Economists also cast the problem in terms of the relative distribution of costs and benefits of alternative waste sites to society in general and the potential mismatch between cost and benefits for specific sectors of society directly impacted by the facilities. In terms of criteria employed by the technical and economic points of view, centralized facilities tended to emerge as the more optimal solution (though specific exceptions may have existed). These two approaches left a considerable gap in the ability to predict the public acceptability of hazardous waste sites for two reasons. First, an underlying uncertainty in the findings of quantitative analyses has persisted, even when the best available techniques and data are used. Second, even given the best technical and economic analyses, the need for a judgment as to which alternative levels of risk to accept has still remained, and such a judgment has been considered beyond the purview of technical analysis.

As a means of compensating for these deficiencies, the approach to the public acceptance of alternative hazardous waste management policies from the political, social, and psychological points of view has become popular as interest groups have become major political forces in directing hazardous waste management policies.[9] Understanding these underlying social and psychological forces is at least as important in developing and implementing hazardous waste policy as the economic and technical considerations. It is from this relatively newer perspective of the social-psychological forces, that this chapter views the question of whether smaller is better when it comes to hazardous waste management.

ADVANTAGES AND DISADVANTAGES OF ONSITE AND OFFSITE APPROACHES TO HAZARDOUS WASTE MANAGEMENT

The Centralized, Offsite Approach

Decisions pertaining to whether waste management should be undertaken at the site of waste generation or at a centralized location have been a problem for a long time.[10] Arguments for centralization have been voiced very strongly, not only for waste management facilities but for potentially hazardous facilities of any kind. This is a peculiarity of our society, which has been driven on the one hand by the need to decentralize its living accommodations, yet centralize many of its services.

The argument for centralization was originally made in terms of economies of scale, yet now has been justified in terms of other criteria, such as safety. For example, in the nuclear area, Wildavsky gives a typical argument for centralization from the safety point of view:

> Experience reveals that single reactors alone on a site are more likely to break down than reactors on sites with multiple units. "A reactor seems to 'learn' significantly faster," Alvin Weinberg notes, "when it is next to 'older brothers,' than when it is an 'only child'." This family aspect of safety extends also to consumers (users) as well as producers of products (doers).[11]

This philosophy is reflected in the actual trends over a couple of decades in the siting of nuclear power plant units. Okrent's list of nuclear power plants that held construction permits by the end of 1979 includes nine units at locations where previous units existed out of a total of forty-eight nuclear power plant locations in 1979.[12]

Centralized treatment and disposal facilities can have the advantage of reducing risk by exposing fewer people to the risks. Under this argument, while waste quantities are larger at a centralized facility, the total number of people at risk may not be much larger than the population exposed at a single onsite facility. In fact, the population is usually very much smaller than the combined total number of people exposed at all individual sites if the wastes were not centralized. In addition, centralization almost always is more economical, since it enjoys economies of scale. Whether these economies would still occur if all transport costs were added in is uncertain.

In the area of hazardous waste treatment, storage, and disposal, about five hundred centralized facilities, operate commercially in the United States, receiving wastes from other sites.[13] Only somewhat over one hundred of these currently have permit applications submitted under RCRA that have been acted upon (either granted or denied) by the EPA. These five hundred facilities account for only about one-quarter of the total number of hazardous waste treatment, storage, and disposal facilities that have submitted RCRA applications (Part B permit applications). Not included in this total are a number of other offsite facilities, called company captive facilities, that receive wastes from various plants within a given company. These are quasi-offsite or centralized facilities, since they do involve some waste transport from other locations.

While quite a number of centralized facilities exist that treat wastes from other locations, new sitings of these large, centrally located commercial facilities, however, have rarely been successful. This outcome has occurred in spite of the acknowledged scale of the hazardous waste problem. Few facilities have been sited also in spite of the variety of approaches taken under an estimated three dozen state siting programs to strengthen state decision making (Daly and Vitiliano, 1987, p. 9). These approaches include state preemption strategies, state override of local decisions, local veto, local control with state procedural restraints (Andrews, 1988), and combinations of these options with programs involving or discouraging active citizen involvement. The limited success of this alternative is reflected in the low number of successfully permitted new facilities. This is exemplified, first, by the fact that most of the recent permits issued for TSDs are for noncommercial facilities. Second, a 1987 national survey of commercial hazardous waste treatment or disposal facilities under state agency jurisdiction estimated that only 81 applications had been received since 1980 by state siting agencies which had legislative authorization to issue permits for such facilities. Of these 81 applications, only six were fully permitted and are now operational and only another two have been approved and were ready for construction as of 1987. Most significantly, the inventory reported that 31 applications were withdrawn or denied; originally there were 14 permits issued but 8 were withdrawn by applicants or were under judicial review (Daly and Vitiliano, June 1987, p. 23). Other states have issued moratoria on the siting of such facilities (Daly and Vitiliano, June 1987, p. 29). One of the few success stories was the hazardous waste processing facility sited in Greensboro, North Carolina by the ECOFLO company, and its success was attributed largely to how the company interacted with the community.[14] According to the 1987 survey, the few successes were very much a function of type of facility: only 2 out of 21 landfills were permitted, 8 out of 26 treatment facilities and 4 out of 24 incinerators.

Various techniques to increase public acceptance of new centralized systems have met with limited success. Educational campaigns, while a critical part of any public involvement program, are not sufficient by themselves to ensure public acceptance, since education often does not consistently influence behavior and risk acceptability.[15] Strategies based on risk perception still require perceptions and behavioral change to be related to one another, in order to use perception as a way of changing behavior. Strategies based on negotiation are highly conditional and only work in very specific situations.[16] A number of other approaches appear under the rubric of risk communication or public participation. Once again, these appear to be useful only at the margin, under certain conditions, and are often not broadly applicable.

Table 11-1 shows the extent to which permits have been issued, primarily at existing facilities, for hazardous waste storage, treatment, and disposal between 1985 and 1988 (excluding federal facilities). This time period reflects actions subsequent to the passage of the HSWA of 1984. The table gives the figures in terms of both numbers of permits issued and facilities to which permits were issued. The total number of permits issued over the four year periods shows that, regardless of whether permits or facilities are counted, non-commercial (onsite, individualized, or decentralized) facilities account for close to three-quarters of the total. The majority of these permits issued were for facilities on existing sites rather than new sites.

The Decentralized, Onsite or Individualized Approach

Treatment and disposal at the site of waste generation has gained indirect and unnoticed public attention over the years as an approach to the waste problem, and is emerging as a significant alternative to new sitings of centralized facilities.

An advantage attributed to onsite treatment and disposal is that only an actual or perceived incremental increase in risk is produced above

Table 11-1.
RCRA Permits Issued to Commmercial and Noncommercial Facilities (nonfederal), 1985-1988

	*Noncommercial**	*Commercial*	*Total*
	A. Number of Facilities		
1985	51	20	71
1986	78	23	101
1987	88	29	117
1988	74	23	97
1985–88 Total	291	95	386
1985–88 Percentage	75.4	24.6	100.0
	B. Number of Permits		
1985	82	40	122
1986	151	44	195
1987	187	68	255
1988	127	54	181
1985–88 Total	547	206	753
1985–88 Percentage	72.6	27.4	100.0

*Includes company captive facilities, which are facilities that serve several plants within a company. These services can involve waste transport from plant to plant.

Source: Computed from U.S. EPA, "B0625 Report," 9/9/88. The classification of commercial facilities is taken from the B0625 report as well.

what is already at the location. It almost never involves a new site for a facility, since it is an integral part of an existing process already in place or at most is a new facility within the property boundaries of an existing facility.[17] Localized operations reduce costs and safety concerns associated with transport to centralized waste management sites.[18]

Onsite facilities are not without their own disadvantages, which include space constraints, environmental conditions at the site, and lower cost-effectiveness relative to centralized facilities. To a large extent the viability and attractiveness of onsite disposal from a legal point of view depends on how liable a waste generator is for his wastes. Stever, Jr. has pointed out that prior to SARA, the waste generator was not liable under Section 7003 for wastes disposed of offsite, making the onsite alternative less attractive. According to Stever, this liability exemption was eliminated under SARA, and a court case after SARA reaffirmed the liability of nonnegligent generators disposing of wastes offsite (*U.S. v. Northeastern Pharmaceutical and Chemical Co.*) (Stever, Jr., 1987, Release #3 3/88, pp. 6–30, 6–31). This would appear to be an incentive for onsite disposal. Also, some have argued that entombment onsite may not have the same degree of liability associated with it as offsite treatment and disposal (Fitzpatrick, Timmerman, and Buelt, June 21–26, 1987, p. 9). Similar arguments apply to how insurable onsite treatment is.

Indications are that in terms of weight of hazardous substances processed, the more popular option by far is onsite treatment and disposal. A 1984 survey by the EPA reported a total of 48,875 large quantity generators in the inventory at that time. Of that total, while 84 percent of the number of generators shipped hazardous wastes offsite for treatment, storage, and disposal, 96 percent by weight of the hazardous wastes were handled onsite (U.S. General Accounting Office (GAO), February 22, 1985, p. 2).

Technologies for Onsite Hazardous Waste Management

The newest technologies for the destruction of hazardous waste often involve high temperature processes and are often portable for use at the site of waste generation. They are able to be conducted in batch operations, which reduce the scale of the facility and increase portability.[19] A variety of thermal destruction techniques have been adapted to onsite waste treatment and are now commercially available. These systems appear to be adaptable to soils contaminated from spills or waste disposal, and many have reported operating results within regulatory requirements. According to the EPA there are five types of mobile incinerators available—the rotary kiln, circulating bed, infrared, fluidized bed, and wet air incinerators (U.S. Congress, Office of Technology Assessment, June 1988, p. 22). Some examples of the first three types of incinerators on the market in 1987 and earlier are described below.

(1) A more conventional modular, mobile system consisting of a rotary kiln, liquid incinerator and heat recovery and pollution control systems was used in a test situation to destroy PCBs and dioxins in Arkansas (Acharya and Reiter, 1987). This was developed by EPAs Office of Research and Development.

(2) A "Circulating Bed Combustor" developed by GA Technologies is based on advanced fluidized bed technology. It is available in a size that can be transported in modules on public roads (White, Yip, and Diot, 1987).

(3) A mobile infrared incinerator uses infrared energy for thermal destruction. The incinerator was first used in June of 1986 for the destruction of PCB in contaminated soil at a Superfund site in Florida (Hill, 1987). The Shirco Company of Dallas, Texas has used an infrared energy unit on a 45 foot long trailer to treat dioxin. It can attain 1,000 C.

(4) A vitrification unit has been developed that can fit into three trailer mounted units including the accommodation of pollution control equipment. It uses electrodes and captures gases under a portable hood (Fitzpatrick, Timmerman, and Buelt, 1987).

Examples of other technologies that are mobile and can be used for onsite waste treatment are also noteworthy:

(1) A low temperature thermal processing system consisting of a heated screw conveyor and afterburner was used on a pilot scale to remove VOCs from soils. The equipment was fully transportable on one flat bed truck (Noland, McDevitt, and Koltuniak, 1987).

(2) A mobile version of an Advanced Electric Reactor using intense thermal radiation and attaining heats of 2,200 C. to 2,800 C. was used to destroy dioxin in contaminated soil in Missouri (Olexsey and Freeman, 1987).

(3) The plasma arc system developed by Westinghouse is available as a mobile unit, and has been used to treat liquid chlorinated wastes. It can attain 10,000 C. (Olexsey and Freeman, 1987, p. 62). At Love Canal a plasma arc high temperature destruction process was used to destroy wastes onsite (Zimmerman, 1987, p. 136).

It is important to underscore the fact that a number of these technologies are still undergoing tests for feasibility and performance. Site demonstrations are usually required prior to their use at a particular waste site.

The Waste Transport Issue as Tipping the Balance

Transport risks have been at the center of the controversy surrounding centralized facilities. This is particularly true in the hazardous waste area, primarily with regard to long-hauling of raw wastes rather than treated wastes. For example, residents of the City of Philadelphia protested vehemently against the transport of wastes through their City for ocean incineration.[20] A similar protest occurred in Brownsville, Texas a few years earlier. The routing of nuclear wastes through concentrated population areas has often been opposed.[21] The transport problem is the key problem for large, centralized waste treatment facilities, not only from the point of view of economics and risk assessment, but from the social-psychological perspective as well. One only has to look at the record of spills of virgin chemicals and oil during transport over the last couple of decades to understand the public sensitivity toward the transport risks associated with chemicals in general. Many of the spills required temporary and occasionally permanent evacuations.[22]

Transport risks have been considered so central to industrial operations that they often have a radical effect on production. For example, after Bhopal, Union Carbide decided to consolidate one of its operations at Institute, West Virginia in order to avoid shipping methyl isocyanate (though that shift in operations eventually created other problems):

> When Union Carbide reopened its MIC unit in May it decided not to ship MIC to its affiliates, but instead to ship the less toxic aldicarb, which is made from aldicarb oxime and MIC (Franklin, August 12, 1985, p. A12).

THE EMERGING POSITION OF GOVERNMENT AND INTEREST GROUPS IN THE DEBATE

Public support by government, special interest groups, and the general public for the onsite waste treatment option has been sporadic but persistent. There are a few powerful examples of such support.

At all levels of government, policies have been formulated to discourage the movement of wastes from their origins. These appear as prohibitions against intercounty and interstate waste transport and more recently, as prohibitions internationally against importation of hazardous wastes.[23] Rate structures for waste processing often attempt to reinforce these policies. For example, processing rates within some states are often less than rates for wastes outside. In Oklahoma, insurance rates were set lower for operators bringing wastes from within the state than for those bringing wastes from outside. This was overturned in a court case, however, because it violated rules for interstate commerce. A recent study summarizing many of the state solid waste legislation pointed out that Ohio's 1988 law allows higher tipping fees at landfills for out-of-state wastes than for wastes originating within the state, but the law has been challenged by the National Solid Waste Management Association for violation of interstate commerce laws (Darcey and Johnson, 1989, p. 24). Other states where similar restrictions have been attempted include Mississippi, Arkansas, North Dakota, and South Carolina. The objections raised on interstate commerce arguments are being argued on several grounds, first whether interstate commerce laws are violated and second, whether a state's jurisdiction over the permit for a landfill extends to controls over where wastes come from (Darcey and Johnson, 1989).

The U.S. EPA is now committed to a policy of onsite, *in situ* cleanup for Superfund sites, under its Superfund Innovative Technology Evaluation (S.I.T.E.) program. A basic part of the Superfund strategy usually involves the construction of an onsite treatment facility. In FY 1987, the U.S. EPA reported that onsite containment was chosen as a remediation alternative for 27 sites out of 75 for which a Records-of-Decision (ROD) had been signed under Superfund (U.S. EPA, Office of Solid Waste and Emergency Response, November 1987, pp. 1–3).

In its oversight review of the Superfund program, the Congress' Office of Technology Assessment has considered onsite treatment options as long as they are permanent and the technology is shown to be effective in meeting environmental standards (U.S. Congress, Office of Technology Assessment, June 1988). Many of the successes that the OTA pointed to in the Superfund program were cases in which the chosen option was an onsite treatment and disposal system.

A coalition of environmental groups commenting on EPA's Superfund program, argued in connection with the choice of containment vs. onsite incineration alternatives that "Even where EPA does not deliberately inflate the cost [of onsite incineration], the data that the Agency is using does not reflect the highly competitive nature of the remedial onsite treatment marketplace" (Environmental Defense Fund et al., June 20, 1988, p. 27). While the report strongly came out against onsite containment as a permanent remedy where environmental conditions were not suited to such containment, the report generally advocated onsite combustion alternatives.

PUBLIC ACCEPTANCE OF HAZARDOUS WASTE FACILITIES

Social-Psychological Criteria for Public Acceptance

Empirical and conceptual studies in behavioral psychology have related certain characteristics of events with the public perception of risk connected with those events. Risk perception has also been linked further to risk acceptability. Briefly reviewing those findings, people are more likely to accept risks to which they are exposed voluntarily, that have an immediate health effect rather than a delayed one, that are known to science and to the public, controllable, familiar rather than new, evolve slowly over time (chronic), are not dreaded, and whose effects are not considered severe or fatal (Slovic, Fischhoff, and Lichtenstein, 1980, pp. 194–195). One that can be added to this list is the management of risks by institutions with good records and who enjoy public trust and credibility. Finally, not explicit in these dimensions (though certainly implied in some of them) is the notion of distance. Several surveys have come up with the intuitively obvious finding that acceptability increases with distance from the respondent's residence or sphere of activity.[24] According to the literature, a technology with all of these characteristics theoretically has a better chance of public acceptance within a given social and cultural setting.

The extent to which these risk characteristics are indicators of public acceptability of risk has only been tested on the general public in a limited number of settings,[25] and the breadth of applicability still has to be more firmly established. Empirical tests have shown that there is considerable variability in the ratings assigned to each of the dimensions outlined above, depending on the kind of risk being scored. Nevertheless, these characteristics provide an important framework to conceptualize public acceptance of controversial facilities.

Other dimensions of public acceptance of risks originate from political theory. For example, controversy can expand issues on the public's agenda, where issue expansion is a function of the size of the public involved.[26] According to Cobb and Elder (1972) and O'Brien, Clarke, and Kamieniecki (1984), the size of public involvement in turn is positively associated with the ambiguity of the issue, its social significance (expressed in terms of the number of people affected by a program and the amount of money involved), how long the issue lasts and whether its relevance extends beyond the immediate case, the degree to which the issue is not technically complex or cast in technical terms, the absence of a clear and generally acceptable solution to the problem, the rapidity with which the issue escalates (for example, in the form of a crisis) and receives public attention.

Public Acceptance of Onsite Facilities and Fears Associated with Offsite Facilities

The social-psychological and political perspectives discussed above provide a basis for conceptualizing relative public acceptance of onsite vs. offsite hazardous waste management facilities. When some of the social-psychological criteria are applied, it would appear that the onsite, individualized systems would be more acceptable than centralized ones.

Knowledgeability of the Risks and Newness

One would expect risks at onsite facilities to be more certain and knowable by virtue of wastes being homogeneous relative to the mixtures from several plants that offsite, commercial facilities often accommodate. Commercial facilities first have to disaggregate wastes and ensure that manifests describing the wastes they receive are accurate. The mix of wastes they have to handle adds an additional element of uncertainty and increases perceived ambiguity to hazardous waste management that is not experienced by an individual facility processing its own wastes.

With regard to the dimension of newness, onsite facilities are not new to the public, since in most cases, they are associated with facilities that have been in a community a long time. In contrast, the commercial facility changes its waste streams constantly, or at least more frequently than a single facility does.

Controllability

The public generally objects to facilities whose geographic scope of operation is very broad and diverse, since such facilities cannot easily be controlled by local, political processes and individual authority. This type of dissatisfaction is often argued by the public in terms of the imbalance between personal or community benefits and the costs they would experience due to the perceived large scale of the facility. In terms of social-psychological criteria, the community begins to feel that they lose control over the facility when its scope becomes enlarged. Morell and Magorian cite a number of hazardous waste management cases that demonstrate this point:[27]

- The Town of Sterling, New York, objected to a hazardous waste facility siting because it felt the facility would end up servicing the entire state's chemical wastes.
- Residents of Westford, Massachusetts, prevented the siting of a waste facility on a granite quarry fearing that the facility would eventually serve all of New England.
- Starr County, Texas citizens opposed a new waste facility on the basis that it would receive wastes from the Houston-Galveston area.
- Residents of Ponce, Puerto Rico, objected to a siting for fear of becoming a dumping ground for U.S. wastes.

Other examples of public objections to the large scope of proposed facilities exist in areas other than hazardous waste management. For example (Zimmerman, 1987, p. 134):

- Public concerns over a second nuclear waste repository were that while a single repository could accommodate existing nuclear wastes, a second repository could result in the expansion of nuclear power plants, an unintended consequence. Thus, limitations on the future expansion in nuclear wastes were not assured, and this perceived threat entered into the debate.
- Local opposition to a power project in Prattsville, New York, proposed by the New York State Power Authority, in part hinged on the fact that Prattsville residents felt that New York City would get the power generated by the upstate facility. The project has now been abandoned by the Authority.
- When limited landings for the Concorde were proposed for JFK and Dulles airports, the public felt that eventually a whole fleet of Concordes (a critical factor in quantifying its impact on stratospheric ozone depletion and hence, human exposure to more radiation) or larger Supersonic Transports (SSTs) might be proposed. The FAA did not limit the scope of Concorde flights as a condition for its approval of limited landings (U.S. DOT, FAA, September 1975, pp. X-1; X-55).

EVALUATION OF RELATIVE ACCEPTABILITY OF CENTRALIZED AND DECENTRALIZED FACILITIES

In order to investigate the degree of public opposition to onsite hazardous waste management facilities and compare this with opposition to centralized (or offsite) alternatives to hazardous waste management, a subset of onsite and offsite facilities, regulated under RCRA and CERCLA (or Superfund), has been identified. Indicators of public acceptance or controversy have been defined in several ways and applied to each subset. First, the results of a relatively comprehensive examination of RCRA and CERCLA or Superfund sites are given applying certain indicators of controversy to these sites on a systematic basis. Second, some specific examples are presented, primarily from the CERCLA or Superfund program, to clarify how acceptable centralized and decentralized waste facilities are.

The Survey

Defining a Subset of Hazardous Waste Facilities

In order to examine the public acceptance of onsite treatment and centralized treatment, a set of hazardous waste management facilities has to be identified that could be divided into each of the two categories. These facilities were discussed earlier as hazardous waste facilities regulated under both the federal RCRA and Superfund programs. While existing inventories do not divide the facilities according to onsite vs. offsite treatment and disposal, one can obtain such a division.

For RCRA facilities, commercial hazardous waste management facilities can be separated out using EPA's inventory of commercial facilities (U.S. EPA, August 1987), thereby obtaining a division into commercial and noncommercial (onsite) facilities. The data in Tables 11-1 and 11-2 give the number of onsite and offsite or commercial facilities that are under the RCRA program.[28] Based on a comparison of the inventory of commercial facilities and RCRA facilities, at least 365 of the facilities that have applied for RCRA permits store and treat their wastes onsite.[29] An unestimatable additional number of "small generators" may be managing hazardous wastes onsite as well. These facilities are defined by the EPA as those that generate less than 100 kg. per month of hazardous wastes.

For Superfund sites, the division into onsite and offsite or centralized facilities is more difficult, and takes on a somewhat different meaning. One has to obtain information on which sites were advanced enough along the path to cleanup to have had Records of Decision (RODs) prepared. The ROD describes the method of treatment and disposal to be selected. Furthermore, the RODs have to be detailed and final enough to specify whether onsite cleanup vs. offsite removal and treatment was used. The OTA has referred to the ROD as the crucial step in the selection of a cleanup technology.[30] Onsite facilities under the Superfund program are those at which the cleanup option involved a permanent, onsite treatment system. By comparison, offsite or centralized treatment relies upon a centralized system and involves waste transport away from the site.[31]

Of the 1,165 sites listed on the National Priorities List as of March 1989, RODs had been prepared for fewer than one-third of the sites. Table 11-3 gives examples of a number of Superfund sites at which onsite treatment and/or disposal technologies were defined in the RODs as distinct from offsite treatment.

Table 11-2.
An Estimate of Offsite and Onsite Facilities in the RCRA Program

Onsite and Offsite RCRA Regulated Facilities for which Permit Decisions have been made by Status and Type of Permit, as of September 1988[1]

	Total Facilities	*Federal Facilities*	*Commercial (Offsite) Facilities*[2]	*Onsite Facilities (nonfederal)*[3]
Storage and treatment				
permit issued	503	46	108	349
permit denied	24	0	8	16
Disposal				
permit issued	128	9	16	103
permit denied	73	1	3	69
Incinerator				
permit issued	62	3	8	51
permit denied	9	2	1	6

Notes:

1. This table gives offsite (or commercial) and onsite or individual hazardous waste treatment facilities by their status under the RCRA program. The facilities listed are only those upon whose permit applications EPA has acted (i.e., granted the permit or denied the permit) as of September 1988. There are actually close to 500 commercial facilities in the EPA inventory, however, permit applications have been submitted for only about 400 commercial facilities, less than 300 commercial facilities have had their permits acted upon (granted or denied) by EPA.

   Note that these figures cannot be totalled, since a given facility may be in more than one permit category. For example, the same facility may hold a RCRA storage and treatment permit, a disposal permit and an incinerator permit.

2. Commercial facilities were those facilities listed in the U.S. EPA, "Directory of Commercial Hazardous Waste Management Facilities" (Springfield, VA: NTIS, August 1987) that also appeared on the September 1988 RCRA list. It is important to note that close to 500 facilities appeared in the commercial directory, whereas probably closer to one hundred or more of those did not appear on the RCRA permit list (the number cannot be calculated exactly because of the overlapping nature of the RCRA status categories). Some (but not all) of the commercial facilities with no RCRA status are Superfund sites.
3. The onsite facilities are estimated by subtracting out commercial or federal facilities from the total RCRA facilities in each category. These also include company captive facilities, which are waste treatment or disposal facilities that receive wastes from other plants within a single company.

Measures or Indicators of Controversy

Once the set of centralized and decentralized facilities is defined for RCRA and Superfund sites, the next step is to define the level of controversy within the two groups. A popular way of operationalizing the concept of public controversy is through media coverage. Studies of the role of the media in attitude formation attest to its importance in

Table 11-3.
Estimates of Onsite and Offsite Hazardous Waste Treatment Under the Superfund Program

Onsite Cleanup*	64
Offsite Cleanup	71
Cleanup Not Yet Proposed/Cleanup Status Unknown or not Determined	216
Total of Superfund Sites with RODs (as of 12/28/88)**	351
Superfund Sites with No RODs	814
GRAND TOTAL	1165

Notes:

*Tabulated from Table 11-5.

**These numbers are in terms of number of sites, not number of RODs. The total number of RODs issued through 12/28/88 was 433 (a number of sites had more than one ROD issued).

either generating public controversy, reflecting actual or latent controversy, or placing public issues on the political agenda, though admittedly it has been difficult to know whether the media creates or reflects public attitudes (Goodell, 1986, p. 386). A study by Mazur (1981), also referenced by Goodell (1986: 588–589), summarized the relationship of the quantity of media coverage and public controversy in the following way: the greater the quantity of media coverage of a controversial issue, the more public opposition to the issue there is likely to be. Davis and Lester observe the force of the media as a factor in leading to public pressure (Davis and Lester, 1988, p. 10).

Along the lines of Mazur's findings and underscored by others, the amount of newspaper coverage of centralized and decentralized facilities was used as an indicator of public controversy over these facilities. This approach is consistent with a tradition of using news media coverage of events, including environmental incidents and accidents, to trace the evolution of an event and the public's attention to and interest in it. Important qualifications, however, are in order. Studies of the quantitative relationship between media coverage and public attention and opposition show that a precise correspondence between media coverage and social perceptions may be limited for several reasons: a considerable amount of filtering goes on in the news media, in the direction of amplifying certain events (Combs and Slovic, 1979) and eliminating coverage of other events (Goodell, 1986). One implication for this analysis is that while the existence of news coverage of an event is likely to signal public interest in the event (enhance or reflect such interest), the absence of coverage does not necessarily imply lack of public interest. In cases where coverage was absent, supplemental indications of public interest and opposition are in order. Another implication is that extensive coverage may not always correspond to heightened public interest, though it probably indicates some level of interest.

Newspaper coverage was confined to the National News Index coverage between 1979 and 1988.[32] The National News Index covers five newspapers with national or at least, broad regional audiences—the *New York Times,* the *Washington Post,* the *Wall Street Journal,* the *Christian Science Monitor,* and the *Los Angeles Times.* Nightly News coverage was also available for about a dozen toxic waste incidents (Greenberg, Sachsman, Sandman, Salomoni, 1989). In a number of cases, media coverage was supplemented by more direct information on the existence of controversy, obtained by in-depth case studies and government records.

The correspondence between the extent of newspaper coverage and controversy depends upon the tone of the articles, and whether they are positive or negative. Supporting the fact that such a correspondence existed in the case of the media coverage of these sites, is first, the general observation that media coverage of scientific issues has become more oriented toward controversy.[33] Second, a scan of the titles and coverage of the articles dealing with RCRA and Superfund sites revealed that articles generally emphasized the negative aspects of the sites.

Survey Results

Prevalence of Onsite Facilities. Onsite RCRA and Superfund treatment and disposal facilities are significantly prevalent in the hazardous waste management service industry. It is clear from the data presented in Table 11-1 that there are quite a large number of facilities in the United States currently engaged in onsite hazardous waste treatment, storage, and disposal that hold permits under RCRA. In fact, these onsite facilities account for about three-quarters of all such facilities in the country. Table 11-2 gives a somewhat different set of such facilities (over a longer time period and with overlapping permit statuses), but reveals a similar pattern of the popularity of onsite facilities over commercial facilities permitted under RCRA. Even when the complicating factor of company captive facilities is taken into account, the number of onsite facilities is quite large. Under the Superfund program, it is relatively more difficult to draw analogous conclusions, given the large number of sites whose cleanup status is as yet undetermined. Table 11-3 gives the number of sites at which cleanup was done onsite vs. offsite. The number is about equal, but the number of sites whose status with respect to onsite or offsite facilities is unknown overwhelms both categories. The main point, however, is that onsite cleanup is so far not an insignificant choice under Superfund and accounts currently for almost one-fifth of the sites that have signed RODs.

Degree of Public Controversy over Onsite and Offsite Facilities. Controversy was measured in several ways. First, a rough indicator of controversy was the number of news entries in several national newspapers indexed by the National News Index (NNI). This approach is generally consistent with the Mazur (1981) hypothesis. A summary of the results of the coverage of cases in the news media for RCRA and Superfund sites is given in Table 11-4.

To the extent that a correspondence between media coverage and public opposition or controversy can be assumed to exist, the following observations emerge:

1. RCRA facilities engaged in storage and treatment, which are generally long-established, operating facilities (regardless of whether they are centralized or decentralized) are far less controversial than Superfund sites—the majority of the RCRA facilities had no NNI entries.
2. Within the category of RCRA facilities studied, there is a slight but probably not significant difference between the percentage of commercial (offsite, centralized) facilities (2%) and the percentage of onsite facilities (7%) that are controversial, i.e., have NNI entries. The dirth of coverage in general obscures the significance of the little difference that does appear. The onsite facilities that are controversial (do have NNI entries) tend to be the very large chemical companies. These are probably company captive facilities, which are quasi-onsite facilities.
3. Within the category of Superfund sites, while sites where onsite treatment is being used appeared more often in the NNI index than those that opted for transfer of material offsite for treatment and disposal, it is critical to point out that the NNI entries occurred prior to the selection of a treatment system. That is, controversy

Table 11-4.
Tabulation of Superfund and RCRA Sites with Respect to News Media Attention

A. Superfund Sites

	Number of sites (RODs)	*Number of Sites with Entries in NNI, 1979–1988* 0	*1–5*	*6–10*	*>10*	*Unknown*
All Superfund Sites with RODs (as of 12/28/88)	351	not available				
Onsite Cleanup	64*	43	(11**)	2	(5**)	3
Offsite Cleanup	71	63***	6	0	2	0
Cleanup Not yet Proposed or Cleanup Status Unknown	216	not available				

B. Selected RCRA-Regulated Sites (Storage and Treatment Only)

	Number of Sites	*Number of Sites with Entries in NNI, 1979–1988* 0	*1–5*	*6–10*	*>10*	*Unknown*
RCRA Sites Storage and Treatment Permits Issued or Denied (non-federal)	481	455	25	1	0	
Onsite (Individual) Facilities	365	341	23****	1	0	
Commercial (Off-site Treatment Facilities)	116	114	2	0	0	

Notes:

*Tabulated from Table 11-5.

**For the majority of these sites, no newspaper items appeared in the NNI after the ROD was signed, indicating an absence of controversy over the selected onsite remedy (see text for discussion).

***Includes many wellfields, which were considered offsite cleanups because they involved finding new water supplied for the communities affected.

****Most of the 23 facilities are the large chemical companies, which are likely to be company captive sites or sites that treat wastes from other plants within the company, transported to a central site. They were in the news for general non-RCRA specific environmental problems also.

drove a solution for the sites, but once the onsite option was chosen, no media attention occurred with one exception. In contrast, while there are fewer NNI entries for offsite treatment, many of these entries occurred even after the offsite option was chosen with only a couple of exceptions.

4. The choice of onsite or offsite options does not appear to be a function of population affected for the Superfund sites. Onsite facilities were chosen in communities whose affected populations ranged from just over one hundred to over one million.

To summarize, the onsite option emerges as often being publicly acceptable. This occurs, first, implicitly because of its widespread use and second, because of the existence of a large number of operating onsite facilities that have not appeared in major national or regional papers over the past ten years. While the relative popularity of operating onsite vs. offsite treatment is not always clear, the onsite option nevertheless is often popular. Some examples of public opinion regarding onsite treatment cases, under Superfund, are discussed below to expand on this finding.

Examples of Public Acceptability of Onsite Hazardous Waste Management Operations

As pointed out earlier, many onsite treatment facilities have been publicly acceptable. In such cases, the public either did not raise objections sufficient to prevent the onsite activity from being undertaken or court decisions did not stop the facility. In cases involving at least implicit acceptability of onsite alternatives, the wastes have been kept at the point of generation for treatment or disposal. The examples below are meant to supplement the analysis of news media coverage via the NNI above, and to begin to gain some insights into the conditions under which the onsite option was considered acceptable. Examples are drawn from the Superfund sites listed in Table 11-5.

Relatively Less Controversial Onsite Treatment Remedies

At the Davis Liquid Waste Site in Smithfield, Rhode Island, the treatment and disposal plan outlined in the Record of Decision (ROD) involved a number of onsite components. It included "onsite thermal destruction of 25,000 cubic yards of excavated raw waste and contaminated soil with greater than 2 parts per million (ppm) of volatile organic chemicals; placement of incineration ash and pollution control residues that are found toxic through testing in an onsite RCRA hazardous waste landfill; . . . and restoration of groundwater by onsite treatment using air stripping and carbon adsorption" (U.S. Congress, OTA, June 1988, p. 5). Part of its public feasibility probably had to do with "the early elimination of nontreatment options" (though "some untreated hazardous material will be landfilled onsite instead of being treated") and the "provision of alternative water for affected offsite residents" (U.S. Congress, OTA, 1988, op cit.). The OTA regarded this solution as a very successful one in the Superfund program, since it involved a permanent solution to the waste problem—a combustion facility. Public concern over onsite incin-

Table 11-5.
Degree of Controversy over Superfund Facilities with Onsite Treatment and Disposal

Ref. No.	Site name	Location	ROD Dates	NNI News Items	Population Affected GW/SW	Air
Sites with a Relatively LOW Degree of Controversy Subsequent to the Choice of an Onsite Facility*						
[1]	Acme Solvent Reclaiming Inc.	Morristown, IL	9/85	0	750	
[3]	A.Y. McDonald Industries	Dubuque, IA		0	62321	
[1]	Bog Creek Farm	Howell Twp., NJ	9/85	0	25514	
[1]	Bridgeport Rental & Oil Services	Bridgeport, NJ	12/84	1**	4859/111	
[7]	Cleve Reber	Sorrento, LA	3/87	0	2500	10000
[1]	Charles George Reclamation Trust Landfill	Tynsborough, MA	9/88	0	2112	
[3]	Chemplex Co.	Clinton/Camanche, IA		0	5774	
[7]	Colbert Landfill	Colbert, WA	9/87	0	1485/809	
[1]	Davie Landfill	Davie, FL	9/85	0	10001	
[4]	Davis Liquid Waste Site	Smithfield, RI	9/85	0		
[1]	Douglassville Disposal	Douglassville, PA	6/88		35000/2855	
[1]	Drake Chemical	Lock Haven, PA	9/84 5/86 9/88	0		10584
[7]	Endicott Village Well Field	Endicott V., NY	9/87	0	45000	
[3]	Fairchild Semi-conductor	South San Jose, CA		3**	65527	
[4]	French Limited	Crosby, TX	3/88	0	5400	
[7]	Geiger (C&M) Oil	Rantoules, SC	6/87	0	900	
[1]	GEMS Landfill	Gloucester Twp., NJ	9/85	3**	38000	38000
[1]	Goose Farm	Plumstead Twp., NJ	9/85	0	0	
[7]	Hardage/Criner	Criner, OK	11/86	0	114	
[7]	Haviland Complex	Hyde Park, NY	9/87	0	2825	
[1]	Harvey & Knott Drum, Inc.	Kirkwood, DE	9/85	0	304	
[1]	Heleva Landfill	North Whitehall Twp., PA	3/85	0	3333	
[1]	Hocomonco Pond	Westborough, MA	9/85	0	12600	
[7]	Katonah Municipal Well	Bedford T., NY	9/87	0	6150	
[7]	Laskin Poplar Oil	Jefferson Twp., OH	8/84 9/87	0	1421	
[7]	Liquid Disposal	Utica, MI	9/87	1**	3576	
[1]	Lone Pine Landfill	Freehold, Twp., NJ	9/84	9	6000/600	1001
[3]	Marley Cooling Tower Co.	Stockton, CA		0	1003	
[2]	Mililani Wells	Oahu, Hawaii		0	19521	
[7]	New Brighton/Arden Hills/ St. Anthony	New Brighton, MN	6/83 9/83	2	95400	
[1]	New Lyme Landfill	New Lyme, OH	9/85	0	1368	

Table 11-5. Continued

Ref. No.	Site name	Location	ROD Dates	NNI News Items	Population Affected GW/SW	Air
[7]	Northside/Environmental Conservation and Chemical Corp.	Zionsville, IN	9/87	0	1000	
[1]	Nyanza Chemical Waste Dump	Ashland, OH	9/85	1	3001/999999	10001
[1]	Old Inger Oil Refinery	Darrow, LA	9/84	0		
[7]	Palmerton Zinc	Palmerton, PA	9/87 6/88	0	2000	
[7]	Palmetto Wood Preserving	Dixiana, SC	9/87	0	5719	
[1]	Picillo Farm	Coventry, RI	9/85	0		
[2]	Piper Aircraft Corp./Vero Beach Water & Sewer Dept.	Vero Beach, FL			33254	
[7]	Renora, Inc.	Edison, Twp., NJ	9/87	0	24000	
[4]	Re-Solve, Inc.	Dartmouth, MA	7/82 9/87		1855	
[7]	Saltville Waste Disposal Ponds	Saltville, VA	6/87	0		1140
[7]	San Fernando Area I	Los Angeles, CA	9/87	1**	999999	
[4]	Seymour Recycling	Seymour, IN	9/86 9/87	3**		
[2]	Sydney Mine Sludge Ponds	Brandon, FL		0	4286	
[1]	Sylvester	Nashua, NH	7/82 9/83	1**	25000	
[7]	Tower Chemical Co.	Clermont, FL	6/87	0	2900	
[1]	Triangle Chemical	Bridge City, TX	6/85	0	4705	
[7]	Volney Landfill	Volney Town, NY	7/87	0	225	
[1]	Whitehouse Waste Oil Pits	Whitehouse, FL	5/85	0	4343	
[7]	Williams Property	Swainton, NJ	9/87	0	4912	
	Sites with a Relatively HIGH Degree of Controversy*					
[4]	Chemical Control	Elizabeth, NJ	9/83 9/87	>10		14258
[4]	Compass Industries	Tulsa, OK	9/87	0		13129
[4]	Conservation Chemical	Kansas City, MO	9/87	3**	114	
[4]	Crystal City Airport	Crystal City, TX	9/87	3**	9820/1040	3126
[5]	Gurley Pit	Edmondson, AK	10/86 9/88	0		
[1]	Hudson River PCBs	Hudson River, NY	9/84	>10	3000	3000
[1]	Lipari Landfill	Pitman, NJ	8/82 9/85 7/88	6	11000/2250	209
[1]	Love Canal	Niagara Falls, NYH	5/85 10/87 9/88	>10**	0/77000	1100

Table 11-5. Continued

Ref. No.	Site name	Location	ROD Dates	NNI News Items	Population Affected GW/SW	Air
[4]	Pristine, Inc.	Reading, OH	12/87	0	13000	
[4]	Sands Springs Petro-chemical Complex	Sand Springs, OK	9/87	0	201	
[4]	Schmaltz Dump	Harrison, WI	8/87	0	330/57000	
[7]	Stringfellow	Glen Avon Hts, CA	7/83 7/84 6/87	>10	40000/16724	16724
[4]	Tacoma Tar Pits (Commencement Bay)	Tacoma, WA	3/88	1**		
[1]	Times Beach	Times Beach, MO	1/84 9/88	>10**	2061	2061

Key to Reference Numbers:

1. U.S. EPA, Hazardous Site Control Division, "Superfund Records of Decision Update." Washington, D.C.: U.S. EPA, various years between 1985 and 1988. These newsletters contain summaries of the RODs, usually indicating whether the cleanup strategy is onsite or offsite. Listings from this source also appeared in reference 7.
2. U.S. EPA, Office of Emergency and Remedial Response, "Descriptions of 149 Sites Proposed for the National Priorities List as of July 1987." HW-8.12. Washington, D.C.: U.S. EPA, July 1987.
3. U.S. EPA, Office of Emergency and Remedial Response, "Description of 43 RCRA Sites Reproposed for the National Priorities List." HW-8.14. Washington, D.C.: U.S. EPA, June 1988.
4. U.S. Congress, Office of Technology Assessment. *Are We Cleaning Up? 10 Superfund Case Studies.* Washington, D.C.: OTA, June 1988.
5. Environmental Defense Fund, et al. *Right Train, Wrong Track: Failed Leadership in the Superfund Cleanup Program.* June 20, 1988.
6. ROD summary in the *Environmental Law Reporter*
7. U.S. Environmental Protection Agency: ROD (Record of Decision) Annual Report, FY 1987. Springfield, VA, NTIS, July 1988. Use of keywords indicating onsite treatment and onsite disposal.

Notes:

*Classification of Sites into Low and High Degree of Controversy. Sites were classified as having a relatively low degree of controversy if (1) they had no entries in the NNI, (2) where entries did exist they were predominantly prior to the signing of the ROD *and* the site was cited in either OTA or the EDF as having had a successful cleanup, or (3) the ROD indicated that there had been no public controversy. Analogously, sites were classified as having a relatively higher degree of controversy if they had entries in the NNI *and* the site was cited in either OTA or the EDF as having had an unsuccessful cleanup.

**Sites whose NNI entries all predate the signing of the latest ROD; Sites with NNI entries, but without any asterisk are those in which some (but a minority) of the NNI entries did appear subsequent to the signing of the latest ROD.

Three of the sites listed also had appeared on Nightly Network News as reported by Greenberg, Sachsman, Sandman and Salomoni (1989, forthcoming). These were: Seymour Recycling (1 time), Love Canal (3 times) and Stringfellow (4 times).

Abbreviations:

GW/SW = Groundwater/Surface Water; NNI = National News Index; ROD = Record of Decision eration was very high, but EPA was apparently able to respond sufficiently to enable the incineration option to proceed (U.S. Congress, OTA, 1988, p. 63).

In Sands Springs, Oklahoma, a cleanup decision was made against incineration and in favor of in-place solidification and entombment. The public opposition against incineration and the decision for solidification was made by local community officials, and the EPA responded to this opposition by endorsing solidification. These officials already constituted an in-place political force that could influence and sustain the decision. The decision was not brought about by an organized environmental group. Another factor that influenced the decision was that people in the town were located fairly far away from the site—according to the EPA hazard ranking system, 4,700 people were affected by groundwater, none by surface water and none by air within the affected radius from the site for each route of exposure.

At the Seymour Recycling Corporation site in Seymour, Indiana, the chosen treatment technology was a combination of a variety of onsite methods. The OTA underscored the thoroughness of the analysis of alternatives and the way that the ultimate selection was made (U.S. Congress, OTA, 1988, p. 7). The site was initially controversial showing three entries in the NNI, but after the ROD was signed, no controversy was apparent through the news media search.

Relatively More Controversial Onsite Treatment Remedies

Several of the Superfund sites at which onsite treatment remedies were selected exhibited controversy measured in terms of the NNI search as well as from other sources independent of the NNI. As shown by the brief discussions below, the controversy surrounded the type of technology chosen and the lack of confidence in its performance, rather than because it was onsite. In fact, in many cases, the alternative supported by the public was onsite.

At the Chemical Control site, which had numerous entries in the NNI (greater than 10), the ROD summary of public concern indicated dissatisfication with the type of technology selected onsite and the way the decision was made, rather than the fact that it was onsite. The community objected because it preferred onsite incineration to capping. The OTA underscored this opposition, but expressed a different concern, noting that *in situ* chemical fixation or stabilization had been chosen without sufficient evaluation.

At the Compass Industries site in Tulsa, Oklahoma, the onsite treatment system selected by EPA was incineration. OTA objected to the fact that the option of capping the landfill had not been evaluated against a broad set of other alternatives, only against incineration. In addition, according to OTA, technical support for the capping option was contradicted by findings at other sites as well as regulatory requirements. The OTA objections apparently were not reflected in public opinion reported in the ROD. Public opinion was in favor of capping and against the possible hazardous emissions posed by incineration (U.S. Congress, OTA, June 1988, p. 27).

The Conservation Chemical site in Kansas City, MO appeared in the NNI three times. The site analysis was regarded by the OTA study as a failure, because of the lack of permanence of the remedy selected (capping and hydraulic containment) and the lack of a comprehensive solution to all of the hazardous wastes on the site. The level of public concern, however, described in the ROD, was considered to be very low (U.S. Congress, OTA, June 1988, p. 33).

At the Crystal City Airport site, much of the controversy surrounded EPA's choice of capping over incineration. OTA and the public opinion cited in the ROD expressed the lack of technical analysis of the onsite incineration option. This was even the subject of a Congressional hearing held at the site on April 11, 1988. As of the writing of the OTA report, various governmental and other organizations were trying, according to OTA, to influence EPA's decision.

An onsite chemical waste treatment facility was constructed at Love Canal even prior to the Superfund program. The ROD dated 10/26/87 proposed the use of an onsite mobile incinerator for the removal and destruction of dioxin (U.S. Congress, OTA, June 1988, p. 6). In spite of the record level of intense public controversy surrounding this site for many years following its discovery, the solution to treat wastes onsite via a new combustion technology ended up being relatively publicly acceptable and was considered by the OTA as one of the more successful Superfund efforts.

At the Pristine, Inc. site in Reading, Ohio, the EPA's selection of *in situ* vitrification met with opposition from the Potentially Responsible Parties (PRPs) and the OTA for poor technical backing for the choice. The ROD reported that the public was divided on the alternative for many different reasons. Nevertheless, the argument was not over whether the treatment alternative should be onsite or offsite, but rather, which onsite alternative was the most suitable.

Taking these examples as being reflective of public opposition to onsite solutions in general, several observations emerge. First, no one treatment and disposal option is ever particularly favored by the public (i.e., incineration, solidification, or capping). The choice does not influence the decision to support the onsite option. Second, the public is more sensitive to the way analyses of sites are done, the way the sites are managed, and other factors described in the next section. Closure of the controversies did occur many times through the selection of onsite remedies, and such closure was reflected in a disappearance of newspaper coverage.

CONDITIONS FOR ACCEPTABILITY OF ONSITE TREATMENT AND DISPOSAL

The concept of an onsite hazardous waste management process will be relatively more acceptable than other forms of hazardous waste management only if certain political conditions are met:[34] Briefly, these conditions pertain to how the benefits of the facility are articulated, the extent to which the community respects the way the facility is operated, public confidence in the technology used, public trust in the management of the facility, and the degree to which the regulatory system supports the management options.

Articulation of Benefits

Community benefits of an onsite facility have to be made clear. As a general statement, support can only occur where a problem has existed before and the treatment is aimed at improving the situation. Public acceptance is usually not as enthusiastic if no treatment is proposed per se, only disposal, i.e., entombment of untreated wastes onsite. The public has often been burned by the fact that entombment has failed to prevent environmental contamination (Zimmerman, 1986). An example of such opposition was state and city opposition to the onsite disposal of thorium wastes at the site of a plant owned by Kerr-McGee in West Chicago, Illinois. The disposal required a permit from the Nuclear Regulatory Commission. The thorium had been on the site prior to its purchase by Kerr-McGee in 1967.[35] A second example is the opposition to onsit disposal of a by-product from the West Valley Demonstration Project process.

Benefits of industrial waste management facilities are not as obvious to the public, since the public is not the generator of the waste. One community benefit of onsite treatment is the absence of the need for waste transport of untreated wastes. Only treated wastes would be transported to another site for ultimate disposal. In line with the issue of benefits, the full environmental tests of effectiveness and efficiency must be demonstrated for the onsite alternatives as well as for any treatment alternative.

Community Respect for Operations

Facility owners and operators must not already be the target of considerable community opposition, i.e., they should be good neighbors. Where existing operations were opposed to begin with, waste treatment operations may be opposed as well. An example is the proposed incinerator siting for the disposal of the U.S. stockpile of chemicals by the Army under the Chemical Stockpile Disposal Program. Destruction of chemical agents and munitions used by the Army has been mandated by September 30, 1994 under P.L. 99-145. After going through the NEPA process, the Army ultimately recommended onsite incineration at eight sites as the preferred alternatives.[36] Communities who had put up with ongoing weapons activities near two of the sites (the Aberdeen Proving Ground in Maryland and the Lexington-Blue Grass Army Depot in Richmond, Kentucky) raised considerable opposition.[37] The nomination of Hanford, Washington as a potential nuclear waste repository site in the mid 1980s was based on the fact that a lot of waste was already there. The Hanford site had developed such a bad reputation for waste leakage[38] that considerable public opposition surrounded its nomination as a repository.

Public Confidence in Technology

There must be public or community confidence in the technology and evaluation procedures for its safety. The effectiveness and safety of the technology used should have been clearly demonstrated. Incineration, which is one of the most popular onsite treatment systems, has received high ratings in terms of effectiveness and reliability by the OTA (U.S. Congress, Office of Technology Assessment, March 1983). OTA also felt that in evaluating Superfund sites, the costs of incineration were often inflated relative to other alternatives. Nevertheless, incineration of hazardous wastes has not been widely accepted or used and onsite incineration has rarely been used on a commercial scale for hazardous wastes. In the eyes of the public this may imply a lack of a track record. In the mid-1980s only 1 percent of all hazardous wastes were incinerated (U.S. Congress, Office of Technology Assessment, August 1986, p. 9). Solidification technologies, in contrast to incineration, received little support from the OTA, primarily because it is not considered a permanent waste destruction technique.

Public Trust in Management

Proposals should be preceded by a sense of public trust in general in the ability of facility operators to manage the technology.[39] Yet, public trust in hazardous waste management has been weak. Events of a decade or more ago still color the public consciousness. Events such as the "midnight dumping"[40] that preceded the passage of RCRA,[41] the illegal chemical spills that were rumored to have followed the oil spill outside of Pittsburgh from the Ashland Oil tank collapse,[42] the illegal dumping of asbestos that followed the issuance of a stringent New York City law,[43] and the illegal dumping of sewage sludge into New York Harbor and adjacent waterways by barges scheduled to bring the wastes to the 106-mile dump site[44] all undermine public confidence in the ability of industry and government to handle wastes voluntarily. The undermining of governmental management of its nuclear weapons manufacturing establishments that precipitated publicly in the press in 1988 has led to further public distrust in the government as a hazardous facility operator. Finally, the history of existing hazardous waste sites is such that the motives of the developers of such sites are highly suspect. Studies have noted that existing hazardous waste sites appear to be sited more often in economically or socially depressed and disadvantaged areas.[45]

Regulatory Support

If onsite treatment is considered a viable option, then the regulatory environment should clearly support it. This is not simple, since government has historically supported offsite, centralized, and large facilities.

These facilities are easier to track and fewer permits have to be issued. The trend, however, may be reversing itself. In 1985, the EPA identified a very active effort by its Office of Research and Development for the development of innovative technologies applicable at least to Superfund sites.[46] The technologies being developed included solidification and stabilization as well as thermal destruction systems. By 1987 the EPA had issued RCRA permits for seven incinerators, one of which was the portable "Circulating Bed Combustor." The unit also received a TSCA permit for destruction of PCBs (White, Yip, and Diot, 1987, p. 13). The critical parameters that determine the acceptability by government of any combustion facility are the emission standards for the facilities and content of ash residue. Many of the portable units for onsite waste treatment have had no trouble meeting the emission standards that currently exist. Ash residues, however, are more problematic, and as of 1988 were the major stumbling block in the permitting of many thermal destruction processes. Local zoning policies often are not consistent with respect to onsite facilities. In fact, many of the battles over these sites occur in the arena of zoning decisions.

CONCLUSIONS

When the public judges how acceptable hazardous waste management strategies are, centralization vs. decentralization of facilities surfaces as a key issue. Facilities decentralized to the site of origin of the wastes are numerous, based on EPA inventories, relative to more centralized facilities. In addition, decentralized, onsite treatment at Superfund sites appears to be more acceptable when it meets public expectations regarding environmental protection. It is difficult to tell whether centralized, commercial facilities currently in operation and regulated under RCRA are acceptable, but clearly new sitings of such facilities have proven to be publicly unacceptable. Because of the prevalence and implicit support occurring for onsite hazardous waste management throughout the country, the balancing of such options against centralized alternatives should be done explicitly.

Onsite waste management systems in many ways pose a radical change in the way hazardous wastes have been managed up until now. The commercial systems are managed by close to sixty companies and encompass over one hundred facilities with federal permits. These commercial systems have handled a lot of the hazardous wastes up until now. There are close to 200,000 hazardous waste generators inventoried in the country that continually make choices as to whether to handle their wastes onsite or offsite. Should onsite systems become more popular, commercial operations would have to shift more into mobile waste reduction technologies if they are to service those sites. All of the management requirements and the public's conditions for acceptability would have to accompany these changes.

ENDNOTES

1. This chapter is an outgrowth of a paper by R. Zimmerman, "A Process Framework for Risk Communication," *Science, Technology, & Human Values,* 12 (1987), 131–137. Portions of this chapter appeared in "Enhancement of Public Acceptability of Hazardous Waste Management through Decentralized (Onsite) Treatment and Isolated Disposal," Zimmerman, R., *Proceedings of the 1988 Hazardous Material Spills Conference,* pp. 770–778 (May 1988). Reproduced by permission of the American Institute of Chemical Engineers (New York, NY), permission request dated January 5, 1989. The research upon which this chapter is based was funded in part from a grant from the New Jersey Institute of Technology's Hazardous Substance Management Research Center and a cooperative agreement from the U.S. Environmental Protection Agency. The author wishes to acknowledge two graduate research assistants at NYU, Penny Miller and Mark Neggers, for their assistance in collecting data upon which this paper is based.
2. 40 *Code of Federal Regulations* Part 264.
3. The lower figure is used by the U.S. Environmental Protection Agency (EPA), the U.S. Congress, Office of Technology Assessment (OTA) and Congressional Budget Office (CBO) and the higher figure was estimated by the Chemical Manufacturers Association (CMA) (CMA, "The CMA Hazardous Waste Survey for 1981 and 1982 (Washington, D.C.: CMA, 1983)) and OTA, "Serious Reduction of Hazardous Wastes" (Washington, D.C.: OTA, September 1986), p. 45.
4. National Academy of Sciences-National Research Council. "Reducing Hazardous Waste Generation. An Evaluation and a Call for Action" (Washington, D.C.: National Academy Press, 1985), p. 9. The U.S. EPA claimed that this was the latest available estimate as of January 1985 and it applied to the wastes generated in the Year 1981 (U.S. Congress, GAO, February 22, 1985).
5. Center for Economics Research, "1986 National Screening Survey of Hazardous Waste Treatment, Storage, Disposal, and Recycling Facilities" (Research Triangle Park, NC: Research Triangle Institute, December 1986), p. 2.
6. OTA. "Ocean Incineration: Its Role in Managing Hazardous Wastes" (Washington, D.C.: U.S. GPO, August 1986), p. 70. Older figures cite a similar growth rate. A report in the mid-1970s put the rate of growth at 5 to 10 percent. The total quantities of nonradioactive industrial hazardous waste cited for the early 1970s only amount to 10 million tons annually (U.S. EPA, "Report to Congress.

Disposal of Hazardous Wastes" (Washington, D.C.: EPA, 1974), p. 3.), compared to the 34.5 million metric tons in 1978 (Council on Environmental Quality. "12th Annual Report" (Washington, D.C.: U.S. GPO, 1982) p. 96) and a 40 million ton figure during the 1980s. This is possibly evidence of growth, but other explanations are better reporting or redefinition of hazardous wastes.

7. These figures were compiled under the EPA Emergency and Remedial Response Information System for action under CERCLA (CEQ, 15th Annual Report (Washington, D.C.: U.S. GPO, 1984), p. 177).
8. Amelia D. Zuras, Francis J. Priznar and C. Scott Parrish, "The National Priorities List Process". In: *Management of Uncontrolled Hazardous Waste Sites,* edited by AlChE (New York, NY: AlChE, 1985), p. 2.
9. Nelkin (1989) has provided an important review of the evolution of the socialpsychological literature as a foundation for risk communication.
10. For an evaluation of waste removal and treatment decisions that predate Superfund and RCRA, see Zimmerman (1986, 1988).
11. A. Wildavsky, *Searching for Safety* (New Brunswick, NJ: Transaction Books, 1988), p. 32. He quotes Alvin Weinberg, "Nuclear Safety and Public Acceptance," *Nuclear News* October 1982), pp. 54–58.
12. These numbers were tabulated from the Tables in Appendix II and III in David Okrent, *Nuclear Reactor Safety* (Madison, WI: U. of Wisconsin Press, 1981), pp. 351–354. In addition, prior to that date, an additional twenty of the forty-eight plants already had at least two units at a given site.
13. This number is based on the U.S. EPA 1987 inventory of such facilities (U.S. EPA, "Directory of Commercial Hazardous Waste Management Facilities" (Springfield, VA: NTIS, August 1987).
14. See F.M. Lynn, "Citizens Involvement in Hazardous Waste Sites: Two North Carolina Success Stories," *Environmental Impact Assessment Review* Vol. 7 (1987), 347–361.
15. For a discussion of this issue and references for it, see Zimmerman (1987), op cit.
16. M. O'Hare, L. Bacow and D. Sanderson. *Facility Siting and Public Opposition* (New York: Van Nostrand Reinhold, 1983), pp. 154–157; J.D. Priscoli, "Conflict Resolution for Water Resource Projects: Using Facilitation and Mediation to Write Section 404 General Permits," *Environmental Impact Assessment Review* Vol. 7 (1987), pp. 321–325.
17. The public generally has had a tradition of rejecting large scale projects. This is true even of relatively nonhazardous facilities, such as sewage treatment plants. An early report by Abt Associates showed that voters typically voted down large wastewater treatment plants and did so particularly where an organized constituency existed (Abt Associates, Inc. "Factors Affecting Pollution Referenda" (Washington, D.C.: U.S. EPA, June 1971)).
18. Even if localization may increase overall treatment costs by not taking advantage of economies of scale, these economies may not matter to the individuals near the site, especially if they are the same people that perceive they will be affected negatively by a larger more centralized facility.

19. Many of the portable combustion facilities, however, share uncertainties that fixed versions have, in spite of the fact that demonstration projects have often produced emissions well within federal Destruction Removal requirements. For example, high temperature combustion leads to various by-products of incomplete combustion when temperatures are too low ("PICs") as well as products of combustion ("POCs"). There are often tradeoffs among chemical emissions at different temperatures. For example, it has been observed that high temperatures are necessary for the destruction of toxic organic compounds; however, this can result in the volatilization of metals, usually minimized at lower temperatures (Visalli, 1987).
20. For a brief account of this case, see M.E. Kraft and R. Kraut, "Citizen Participation and Hazardous Waste Policy Implementation." In: *Dimensions of Hazardous Waste Politics and Policy,* edited by C.E. Davis and J.P. Lester (Westport, CT: Greenwood Press, 1987), pp. 64–65. Also see, L. Gruson, "Philadelphians at Hearing Protest Plan to Store Hazardous Waste for Ocean Burning," *The New York Times,* January 14, 1986, p. 8.
21. See, for instance, R. Smothers, "Route of L.I. Nuclear Waste is Argued at House Hearing", *New York Times* (August 2, 1985).
22. These spills are reported on a consistent basis in the records of the U.S. Department of Transportation's Hazardous Materials Transportation bureau and the National Response Center.
23. A United Nations conference held in March, 1989 resulted in an international treaty involving 34 nations that required countries to determine the adequacy of their treatment capability before hazardous wastes are accepted from other countries. Steven Greenhouse, "U.N. Conference Supports Curbs On Exporting of Hazardous Waste," *New York Times* (March 23, 1989), p. A-1.
24. A Council on Environmental Quality survey in 1980 found that 90 percent of respondents said they would find the location of a hazardous facility 100 miles from their homes acceptable. (CEQ, "11th Annual Report" (Washington, D.C.: U.S. GPO, 1980), p. 415. The exception to this, however, is the intense public concern over ocean disposal and incineration of wastes and off-shore drilling for oil. It is possible that the on-shore effects, which are near to people, are what is creating the opposition, not the distance of the activity itself.
25. The acceptability of risks by experts tends to be consistent with some of the reactions of the general public. For example, Lindell and Earle (1983) showed that risks were more acceptable that were more familiar to professionals than those that weren't. Nuclear scientists feared nuclear wastes less than chemical wastes and chemical engineers had the opposite reaction.
26. O'Brien, Robert M., Michael Clarke, and Sheldon Kamieniecki, "Open and Closed Systems of Decision Making: The Case of Toxic Waste Management," *Public Administration Review* (July/August 1984), pp. 334–340.
27. Morell, D. and C. Magorian. *Siting Hazardous Waste Facilities* (Cambridge, MA: Ballinger, 1982), pp. 25–26. These cases were also cited in Zimmerman (1988), op cit., p. 134.
28. The concept of onsite facilities may be somewhat of a simplification. For

example, hazardous waste treatment facilities owned by an industrial plant may be treating wastes generated at the site where the treatment facility is located and they may be receiving wastes for treatment from plants at other sites but owned by the company. These so-called company captive facilities have been incorporated under the category of onsite facilities, however, the interpretation of the data on controversy has taken the classification problem these facilities pose into account.

29. It is not possible to estimate the rest because of the overlapping nature of the RCRA permit status categories. The degree of overlap, however, is very minor: only seven facilities had both incineration and disposal permit reviewed simultaneously at the same site; only three facilities had an incineration permit and storage and treatment permit reviewed at the same site; and only one facility had a disposal permit and storage and treatment permit reviewed at the same site. This indicates the highly specialized nature of the waste management industry.
30. The OTA report provides a good summary of the decision documents produced once a site is added to the National Priorities List. A Remedial Investigation Feasibility Study (RIFS) starts the process with site investigations to determine the extent of contamination. One or more Records of Decision (RODs) can be produced, describing remedial technologies. These are followed by a Remedial Design (RD) document and finally a Remedial Action (RA) document. The later RODs, the RD and the RA all can provide information on whether cleanup was onsite or offsite.
31. Where combinations of the two options existed, onsite status was assigned only to sites where a recognizable, permanent and ongoing facility was to exist at the site.
32. Later analyses of this data will be broadened to include other newspaper indices.
33. Goodell (1986:587–8) summarizes several surveys of the media's coverage of scientific issues, noting the emphasis upon conflict in the news stories.
34. Many of these conditions were discussed by Zimmerman (1987).
35. "Kerr-McGee is Cutting Troubled Nuclear Role," *New York Times* (January 6, 1986), p. A 12.
36. These recommendations were proposed in its Draft Programmatic Environmental Impact Statement dated July 1, 1986.
37. CEQ, 16th Annual Report, (Washington, D.C.: U.S. GPO, 1987), p. 157. Hearings held at the Kentucky site were held by the House Investigations Subcommittee of the Committee on Armed Services, July 25, 1986.
38. This is partly documented in U.S. Congress, Office of Technology Assessment, "Managing the Nation's Commercial High-Level Radioactive Waste" (Washington, D.C.: U.S. GPO, March 1985), Appendix A-2. In mid-1988, the government was strongly considering not reopening the facility at all.
39. Zimmerman (1987), op cit.
40. 45 Federal Register 33084-5 (1980).
41. Midnight dumps accounted for an estimated 26 percent of waste removal

actions conducted by EPA under Superfund in FY1984–FY1985 alone (U.S. EPA, "Superfund: A Six Year Perspective" (Washington, D.C.: U.S. EPA, October 1986), p. 9.).

42. "U.S. Rebuts Rumors of Poison Dumping," *New York Times* (March 12, 1988).
43. "Illegal Asbestos Dumping Soars Under New Law," *New York Times* (June 10, 1987), p. B1.
44. "Barge Company Accused of Dumping Sewage Sludge," *New York Times*, December 25, 1988.
45. See for example, U.S. General Accounting Office, "Siting of Hazardous Waste Landfills And Their Correlation with Racial and Economic Status of Surrounding Communities" (Washington, D.C.: U.S. GPO, June 1, 1983); Commission for Racial Justice, United Church of Christ. Toxic Wastes and Race in the United States (New York: United Church of Christ, 1987).
46. U.S. EPA, Hazardous Site Control Division, "Superfund Records of Decision Update", Vol. 1, No. 8 (November 26, 1985).

REFERENCES

Abt Associates, Inc.: Factors Affecting Pollution Referenda. Washington, D.C., U.S. EPA, June 1971.

Acharya, P. and Reiter, B.A.: PCB Trial Burn in a Modular, Movable Incinerator. Paper presented at the 80th Annual Meeting of the Air Pollution Control Association, New York, NY, June 21–26, 1987.

Andrews, Richard N.L.: Hazardous Waste Facility Siting: State Approaches. In Davis, Charles E. and James P. Lester (Eds.): *Dimensions of Hazardous Waste Politics and Policy.* Westport, CT, Greenwood, 1988, pp. 117–128.

Barge Company Accused of Dumping Sewage Sludge. *New York Times,* December 25, 1988.

Center for Economics Research: 1986 National Screening Survey of Hazardous Waste Treatment, Storage, Disposal, and Recycling Facilities. Research Triangle Park, NC, Research Triangle Institute, December 1986.

Chemical Manufacturers Association. The CMA Hazardous Waste Survey for 1981 and 1982. Washington, D.C., CMA, 1983.

Cobb, Roger W. and Elder, Charles D.: *Participation in American Politics: The Dynamics of Agenda Building.* Baltimore, MD, Johns Hopkins University Press, 1972.

Combs, Barbara and Slovic, Paul: Newspaper Coverage of Causes of Death. *Journalism Quarterly, 56 (4):* 837–343, 1979.

Commission for Racial Justice, United Church of Christ: Toxic Wastes and Race in the United States. New York, United Church of Christ, 1987.

Council on Environmental Quality: 11th Annual Report. Washington, D.C., U.S. GPO, 1980.

Council on Environmental Quality: 12th Annual Report. Washington, D.C., U.S. GPO, 1982.

Council on Environmental Quality: 15th Annual Report. Washington, D.C., U.S. GPO, 1984.

Council on Environmental Quality: 16th Annual Report. Washington, D.C., U.S. GPO, 1987.

Daly, John B. and Vitaliano, Eric N.: Hazardous Waste Facility Siting. A National Survey. Albany, NY: Legislative Commission on Toxic Substances and Hazardous Wastes, June 1987.

Darcey, Sue and Johnson, Bruce: Long Haul: The Answer to Landfill Shortages? *World Wastes:* 24–25, 52–56, March 1989.

Davis, Charles E. and Lester, James P. Hazardous Waste Politics and the Policy Process. In Davis, Charles E. and James P. Lester (Eds.): *Dimensions of Hazardous Waste Politics and Policy.* Westport, CT, Greenwood, 1988, pp. 1–34.

Douglas, Mary: *Risk Acceptability According to the Social Sciences.* New York, NY, Russell Sage, 1986.

Douglas, Mary and Wildavsky, Aaron: *Risk and Culture.* Berkeley, CA, University of California Press, 1982.

Environmental Defense Fund, et al.: *Right Train, Wrong Track: Failed Leadership in the Superfund Cleanup Program.* Washington, D.C., EDF, June 20, 1988.

Fitzpatrick, V.F., Timmerman, C.L. and Buelt, J.L.: In Situ Vitrification—An Innovative Thermal Treatment Technology. Paper presented at the 80th Annual Meeting of the Air Pollution Control Association, New York, NY, June 21–26, 1987.

Franklin, Ben A.: Toxic Cloud Leaks at Carbide Plant in West Virginia. *New York Times,* August 12, 1985, p. A12.

Goodell, Rae: The role of the mass media in scientific controversy. In Englehardt, H.T., Jr. and Caplan, A.L. (Eds): *Scientific Controversies.* Cambridge, Eng., Cambridge University Press, 1986.

Greenberg, Michael R.; Sachsman, David B.; Sandman, Peter M.; and Salomoni, Kandice L.: Risk, Drama and Geography. Network Evening News Coverage of Environmental Risk. *Journalism Quarterly:* 1989 forthcoming.

Greenhouse, Steven: U.N. Conference Supports Curbs On Exporting of Hazardous Waste. *New York Times,* March 23, 1989, p. A-1.

Gruson, L.: Philadelphians at Hearing Protest Plan to Store Hazardous Waste for Ocean Burning. *The New York Times,* January 14, 1986, p. 8.

Hill, M.: Mobile Infrared Incinerators for Soil Decontamination. Paper presented at the 80th Annual Meeting of the Air Pollution Control Association, New York, NY, June 21–26, 1987.

Illegal Asbestos Dumping Soars Under New Law. *New York Times:* p. B1, June 10, 1987.

Kraft Michael E. and Kraut, Ruth: Citizen Participation and Hazardous Waste Policy Implementation. In Davis, Charles E. and James P. Lester (Eds.): *Dimensions of Hazardous Waste Politics and Policy.* Westport, CT, Greenwood, 1988.

Lindell, Michael K. and Earle, Timothy C.: How Close is Close Enough: Public Perceptions of the Risks of Industrial Facilities. *Risk Analysis, 3:* 245–253, 1983.

Lynn, Frances M.: Citizens Involvement in Hazardous Waste Sites: Two North

Carolina Success Stories. *Environmental Impact Assessment Review 7:* 347–361, 1987.

Mazur, Allan: Media Coverage of Technical Controversies. *Journal of Communication, 31:* 106–115, 1981.

Morell, David and Magorian, Christopher: *Siting Hazardous Waste Facilities.* Cambridge, MA, Ballinger, 1982.

National Academy of Sciences-National Research Council: Reducing Hazardous Waste Generation. An Evaluation and a Call for Action. Washington, D.C., National Academy Press, 1985.

Nelkin, Dorothy: Communicating Technological Risk: the Social Construction of Risk Perception. *Ann. Rev. Public Health 10:* 95–113, 1989.

Noland, J.W., McDevitt, N.P. and Koltuniak, D.L.: Thermal Stripping Treatment of Soils Contaminated with Volatile Organic Compounds. Paper presented at the 80th Annual Meeting of the Air Pollution Control Association, New York, NY, June 21–26, 1987.

O'Brien, Robert M.; Clarke, Michael and Kamieniecki, Sheldon: Open and Closed Systems of Decision Making: The Case of Toxic Waste Management. *Public Administration Review:* 334–340, July/August 1984.

O'Hare, Michael, Bacow, Lawrence and Sanderson, Debra: *Facility Siting and Public Opposition.* New York, NY, Van Nostrand Reinhold, 1983.

Okrent, David: *Nuclear Reactor Safety.* Madison, WI, University of Wisconsin Press, 1981.

Olexsey, R.A. and Freeman, H.M.: Managing Dioxin. *Civil Engineering:* February 1987.

Priscoli, Jerome Delli: Conflict Resolution for Water Resource Projects: Using Facilitation and Mediation to Write Section 404 General Permits. *Environmental Impact Assessment Review, 7:* 321–325, 1987.

Rogers, Everett M. and Sood, Rahul S.: Mass Media Communication and Disasters: A Content Analysis of Media Coverage of the Andhra Pradesh Cyclone and the Sahel Drought. In Committee on Disasters and the Mass Media, National Research Council (Ed.): *Disasters and the Mass Media.* Washington, D.C., National Academy Press, 1980, pp. 139–157.

Slovic, Paul, Fischhoff, Baruch and Lichtenstein, Sarah: Facts and fears: Understanding perceived risk. In Schwing, Richard and W.A. Albers, Jr. (Eds.): *Societal Risk Assessment: How Safe is Safe Enough?.* New York, NY, Plenum, 1980, pp. 194–195.

Smothers, R.: Route of L.I. Nuclear Waste is Argued at House Hearing. *New York Times:* August 2, 1985.

Stever, Donald W., Jr.: *The Law of Chemical Regulation and Hazardous Wastes.* New York, NY, Clark Boardman, 1987.

U.S. Congress, Office of Technology Assessment: *Are We Cleaning Up? 10 Superfund Case Studies.* Washington, D.C., OTA, June 1988.

U.S. Congress, Office of Technology Assessment: Ocean Incineration: Its Role in Managing Hazardous Wastes. Washington, D.C., U.S. GPO, August 1986.

U.S. Congress, Office of Technology Assessment: Serious Reduction of Hazardous Wastes. Washington, D.C., OTA, September 1986.

U.S. Congress, Office of Technology Assessment: Technologies and Management Strategies for Hazardous Waste Control. Washington, D.C., U.S. GPO, March 1983.

U.S. DOT, FAA: Concord Supersonic Transport Aircraft. Final EIS, Vol. 1. Washington, D.C., U.S. DOT, September 1975.

U.S. Environmental Protection Agency: Directory of Commercial Hazardous Waste Management Facilities. Springfield, VA, NTIS, August 1987.

U.S. Environmental Protection Agency: Report to Congress. Disposal of Hazardous Wastes. Washington, D.C., EPA, 1974.

U.S. Environmental Protection Agency: ROD (Record of Decision) Annual Report, FY 1987. Springfield, VA, NTIS, July 1988.

U.S. Environmental Protection Agency: Superfund: A Six Year Perspective. Washington, D.C., U.S. EPA, October 1986.

U.S. Environmental Protection Agency, Office of Emergency and Remedial Response: Descriptions of 149 Sites Proposed for the National Priorities List as of July 1987. HW-8.12. Washington, D.C., U.S. EPA, July 1987.

U.S. Environmental Protection Agency, Office of Emergency and Remedial Response: Descriptions of 43 RCRA Sites Reproposed for the National Priorities List. HW-8.14. Washington, D.C., U.S. EPA, June 1988.

U.S. Environmental Protection Agency, Office of Solid Waste and Emergency Response: Solid and Hazardous Waste Report for Fiscal Year 1987. Washington, D.C.: U.S. EPA, November 1987.

U.S. Environmental Protection Agency, Office of Solid Waste and Emergency Response: B0625 Report. Washington, D.C., U.S. EPA, 9/9/88.

U.S. General Accounting Office: Illegal Disposal of Hazardous Waste: Difficult to Detect or Deter. Washington, D.C., U.S. GAO, February 22, 1985.

U.S. General Accounting Office: Siting of Hazardous Waste Landfills And Their Correlation with Racial and Economic Status of Surrounding Communities. Washington, D.C., U.S. GPO, June 1, 1983.

U.S. Rebuts Rumors of Poison Dumping. *New York Times,* March 12, 1988.

Visalli, Joseph R.: A Comparison of Dioxin, Furan and Combustion Gas Data from Test Programs at 3 MSW Incinerators. *J. of the Air Pollution Control Association, 37:* 1451–1463, December 1987.

Weinberg, Alvin: Nuclear Safety and Public Acceptance. *Nuclear News:* 54–58, October 1982.

White, M.L., Yip, H.H. and Diot, H.R.: Transportable Circulating Bed Combustor for Thermal Treatment of Contaminated Soils and Sludges". Paper presented at the 80th Annual Meeting of the Air Pollution Control Association, New York, NY, June 21–26, 1987.

Wildavsky, Aaron: *Searching for Safety.* New Brunswick, NJ, Transaction Books, 1988.

Zimmerman, Rae: The Management of Risk. In Vincent Covello, Joshua Menkes, and Jeryl Mumpower (Eds.): *Risk Evaluation and Management.* New York, NY, Plenum, 1986.

Zimmerman, Rae: The Management of Risk. Executive Summary, Final report to the National Science Foundation. New York, NY, NYU, 1982.

Zimmerman, Rae: Federal-State Hazardous Waste Management Policy Implementation in the context of Risk Uncertainties. In Charles E. Davis and James P. Lester (Eds.): *Dimensions of Hazardous Waste Politics and Policy.* Westport, CT, Greenwood, 1988, pp. 177–204.

Zimmerman, Rae: A Process Framework for Risk Communication. *Science, Technology, & Human Values, 12:* 131–137, 1987.

Zuras, Amelia D., Priznar Francis J., and Parrish, C. Scott: The National Priorities List Process. In AICHE (Ed.): *Management of Uncontrolled Hazardous Waste Sites.* New York, NY, AIChE, 1985, p. 1–3.

45 Federal Register 33084-5 (1980).

Chapter 12

THE RETURN OF THE REPRESSED: FREUDIAN THEORY, HAZARDOUS WASTE SITING, AND PUBLIC RESISTANCE

RICHARD WALKER[1]

I do not wish to arouse conviction; I wish to stimulate thought and to upset prejudices.

—Sigmund Freud, Lecture XVI, 1917
(1916–1917, 243)

Preface

The purpose of this paper is to apply psychoanalytic theory to the problem of siting a waste[2] facility in the United States in the hope that it will stimulate thought and upset prejudices. Community resistance to hosting all such wastes, from solid to radioactive, is common and shares a similar etiology. The public record in finding host communities for such facilities is consistently dismal. This is true whether the primary siting agent is local, state, or federal government agencies or private consultants. One need only recall a 3,000-ton barge of orphaned solid wastes at sea along the Atlantic seaboard, searching in futility from state to state, then from nation to nation for a host site, to grasp the seriousness and the dimension of the problem. Resistance to siting efforts for all

[1]The analysis and opinions expressed in this chapter represent solely those of the author and do not represent any other person or organization including the author's employer, the State of Illinois. The author is a Senior Policy Analyst in the Office of Nuclear Facility Safety at the Illinois Department of Nuclear Safety in Springfield.

[2]Note: "Hazardous waste" and "waste" are used generically and refer to solid, toxic, chemical, radioactive, or any other form of man-made waste which is dangerous to human or ecological safety. "He," "his," and "him" are also used generically and represent both sexes.

types of waste, whether solid, toxic, or radioactive, is commonly manifest as anxiety, fear, hostility, hysteria, and threats of violence.

The general public has virtually no conception of the characteristics of civilization's nonorganic wastes. Therefore, a person's only frame of reference for reaction, consciously or unconsciously, is personal experience. The task is to discover the association between the way in which we, as culture-bearing individuals, have been taught to think unconsciously about our own personal organic waste, specifically feces and defecation, and how we think about and react to the idea of hosting a waste site. We want to know if the origin of our distaste of the former is the origin of our resistance to the latter.

MORAL, REALISTIC, AND NEUROTIC ANXIETIES

Experience in waste siting or its literature reveals three major components of the problem: (1) the Equity Issue ("Why here? Why not in someone else's back yard?") (Kasperson, ed., 1983); (2) the Safety Issue ("Is this going to ruin the environment?" "Our property values?" "The genes of our grand-children?" "What are the risks?"); and (3) the Trust Issue ("We don't trust you to do your job or to have our interests at heart. You're in it for the money. Why won't you grant us the authority to shut it down?") (Colglazier, 1986).

These three issues reflect anxieties associated with the ego, the superego, or both. Equity Issues are moral issues and create *moral anxiety* to the superego. Safety Issues are essentially objective in nature and create *realistic anxiety* to the ego (Freud, 1923, 46). Trust Issues are both moral and objective and therefore generate a mixture of moral and realistic anxieties.

Efforts at siting unpopular facilities (airports, pipelines, power plants, oil refineries, etc.) and allied efforts at public participation and education have focused on one or more of these three categories of concern across these two levels, ego and superego. One finds in the siting literature a familiar litany which summarizes "the nature of the siting problem": lack of a systems approach, risk uncertainty, public perception of risk, inequity in costs and benefits, institutional distrust, difficulties in risk communication, and the availability of ample institutional means for resisting facility imposition (Kasperson, 1985, 11). One hears repeated the assumption: "Undoubtedly the most prominent concern in siting any hazardous waste facility . . . is the question of risk—in other words, is the facility safe?" (Morell and Magorian, 1982, 37). Again: "The paramount issue in hazardous waste facility siting is undoubtedly the risk to health, safety, and environmental quality" (Sandman, 1987, 333).

Traditional efforts, successful in siting many types of undesirable facilities, have been singularly unsuccessful when it comes to waste facility siting. Thus, Kasperson (1985) poses the fundamental question:

> Why has hazardous waste facility siting run aground while other unwanted facilities continue to find host communities despite local resistance?

There has been no failure of effort, rather a failure of theory. This paper rejects the commonly accepted assumptions voiced above, identifies a hidden dimension of the problem which separates waste siting issues apart from the others, and provides a new answer to Kasperson's piercing question. There is another cluster of issues on a deeper level which is wholly neglected in the literature—the unconscious dimension (id). This is the Waste Issue, or for clarity, "the Waste Itself Issue." It is here, beneath the surface, at the level of the id where one finds the system of unspoken repressed forces which create *neurotic anxiety* (Freud, 1926, 164–168).

In the context of waste siting, the term "neurotic anxiety" has two different meanings. First, neurotic anxiety may mean a phobic reaction to wastes which, like all irrational fear, is out of all proportion to their actual danger (Rycroft, 1968, 7–8). Scientists and risk analysts lament and ridicule public resistance in this sense of the term, as though public misperception of risk alone accounts for much of community resistance. On the contrary, even at this level, the "hysterical contagion" observed in various communities, upon careful sociological analysis, may be understood as rational and justified (Schwartz et al., 1985, 72).

Second, neurotic anxiety may mean a response to an as yet unrecognized factor in the self evoked by the stirrings of unconscious, repressed forces hidden in the self (Rycroft, 1968). This chapter is concerned exclusively with this second, deeper meaning of the term. It shall therefore penetrate far beyond the surficial "risk perception," "risk analysis," and similar techno-rational approaches typically embraced by scientists, social scientists, and the waste managers who hire them.

The thesis is simple. It is the power of these hidden forces and the failure to recognize and appreciate them—indeed, a generally pervasive attempt to avoid or deny them—which has contributed to the dismal record of waste siting projects in the U.S. The question, then, is this:

What is the source of this "system of repressed forces" which causes such neurotic anxieties and which fosters a reluctance to deal with them? Bluntly put, What makes waste different?

A Vertical, Dynamic Theoretical Perspective

There are six advantages of this theoretical perspective over the traditional categorizations: (1) The added dimension of the Waste Itself Issue deepens the perspective. (2) These four concerns—Equity, Trust, Safety, and now Waste Itself—rather than being viewed as conceptually isolated and disconnected, are conceived as hierarchical and interrelated. (3) They are interrelated through the fundamental relationships of superego, ego, and id and hence through their respective types of anxieties, specifically moral, realistic, and neurotic, as noted above. (4) The etiology of these anxieties is traceable to the erotogenic stages of psychosexual development (oral, anal, phallic, and genital). (5) The dynamic of these interrelationships is the well-known defense mechanisms of displacement,[3] reaction formation, and symbolization. (6) Finally, these concepts are applicable at the collective as well as the individual level. We can therefore extend them to a community as a whole (Freud, 1930, 136–144).

Figure 12-1 gives a summary of these hierarchical relationships by category. The id, ego, and superego are, respectively, represented collectively as antisocial egocentrism (especially aggression, violence, or fraud), social structure (societal organizations and institutions), and cultural values (Freud, 1930). The perspective, then, is vertical and dynamic, rather than confined to one horizontal level of analysis (Durkheim, 1895). Before tracing the origins of waste neurosis, the following sections illustrate three defense mechanisms at work in waste management—reaction formation, displacement, and symbolization—and survey other theoretical perspectives.

Reaction Formation to Issues

It is a principle of psychoanalytic theory that when ego fails in its allegiance to the *reality principle* by capitulating to the *pleasure principle*

[3]In psychoanalytic terms, displacement refers to the substitution of one object for another; sublimation refers to the rechannelled energy or emotion which accompanies the displacement. "Displacement" is used here to refer to both. Other psychoanalytic terms used in this paper are defined in Rycroft, 1968.

PSYCHOANALYTIC LEVEL	CATEGORIES Individual	Collective	Waste Issues	Level of Anxiety
SUPEREGO:	CONSCIENCE	CULTURAL VALUES	EQUITY & TRUST	MORAL
EGO:	IDEAS AND BELIEFS	ORGANIZATIONS & INSTITUTIONS	SAFETY & TRUST	REALISTIC
ID:	PASSIONS & DESIRES	AGGRESSION VIOLENCE, FRAUD	WASTE ITSELF	NEUROTIC

Figure 12-1. Correspondence between concepts across psychoanalytic hierarchy of dimensions.

(id), it attempts to maintain its weakened dignity by hiding this truth from a nagging conscience (superego) (Freud, 1911, 223–6; Freud, 1923, 57). Just as children lie in order to avoid censure from their parents, in adulthood, having internalized those authorities, we lie to ourselves (Freud, 1923, 54–6).

Communities and organizations behave the same. When their conduct does not measure up to the noble goals or values which they claim to pursue or represent, they attempt to cover-up their baser motives and activities. Bureaucracies are well-known for the manufacture, propagation, and defense of rosy, self-serving propaganda at the expense of truth (Altheide and Johnson, 1980). The problem is not just that bureaucracies lie, but that they honestly believe their lies. This creates a powerful, structural self-delusion (Gross, 1980, 259).

This "lying" about reality is done by using defense mechanisms (A. Freud, 1937), one of which is reaction formation (Rycroft, 1972, 136). In practice, a reaction formation is a form of denial and deception designed to allow one to "lie honestly" to himself or to others. For examples, solicitude may mask an interest in cruelty, cleanliness may mask an interest in dirt, or politeness may mask hostility. Such motives, however, may be so transparent to others that, as Emerson (1860) put it, "The louder he spoke of his honor, the faster we counted our spoons." At the organizational level, a similar hypocrisy is often apparent. Borrowing from Shakespeare (*King Richard III,* iii, vi), one is compelled to exclaim, "Who is so gross that cannot see this palpable device! Yet who so bold but says he sees it not!"

A principle important to waste siting is that the stronger the temptations to ego, that is, to institutions in a community to violate its conscience or values (superego), the less stable it is, the more anxious it becomes, and the more easily regression to previous forms of behavior take place when difficulties arise. But, also, stronger is the formation of reactions to prevent this from happening and more strident is the mask of publicly presented defensive statements. Therefore, there will be occasions on which various factions in a waste siting process will publicly "lie honestly." A community may show an interest in hosting a waste site, then suddenly disavow it. Or a siting agent may promise that a facility will bring hundreds of jobs to a local community, only to deny it later or claim "plans changed" or "we were misunderstood." Therefore, public comments or pronouncements should never be taken at face value.

In some siting efforts, attitudes in a community may be far worse than they seem; in others, better. There is a sense, then, in which scientifically accurate and objective sources of information, if they do not include unconscious motives, can be both accurate and wrong. That is, objective facts may actually conceal truth about a community, rather than reveal it.

As illustration, consider the early stages of the Wilsonville, Illinois, toxic waste controversy. In response to intense emotions, the town council ordered that a trench be dug across the entrance to the waste site located at the edge of town. This effectively closed it. Later, after the Illinois Appellate Court ruled in the community's favor, the council and citizens denied that the trench had had anything to do with the waste site, but that street drainage in that end of town was acutely in need of repair and a culvert had to be replaced (O'Hare et al., 1983, 22) (The psychoanalytic nature of civil disobedience is shown in Figure 12-2).

Displacement of Issues

One is familiar with the phenomenon of displacement in fitful dreams where an emotion such as fear may be displaced onto wholly innocent objects or persons such as those we love (Freud, 1900). This happens at town meetings. Citizens under stress or in turmoil may speak one language and mean another. This represents a displacement of emotions (id), ideas (ego), or values (superego) onto inappropriate subject matter. This occurs whenever neurotic anxieties seek outlets for expression. All three levels of dialogue can be mixed-and-matched, resulting in strange cross-talk, "venting" at bizarre times, emotional outbursts followed by periods of passivity, sudden bursts of righteous indignation over trivial issues, reversals of positions or logic, and tremendous confusion or obfuscation of issues. Just as dreams show no respect for time or logic and hence appear irrational, so do public meetings.

Meetings often take on the same irrational qualities as dreams because at meetings, as in sleep, the ego—with its twin creations of time and logic—is dissolved. In dreams, however, time is condensed; in meetings time is dilated. Many persons are drawn to mass gatherings precisely because they find satisfaction in the ego-loss which the group provides and the simultaneous opportunities to vent repressed emotions publicly. Groups regress easily to primitive mental activity of a kind Freud (1921, 122) ascribed to the primal horde:

> . . . the dwindling of the conscious individual personality, the focusing of thoughts and feelings in a common direction, the predominance of the affective side of the mind and of unconscious psychical life, the tendency to the immediate carrying out of intentions as they emerge . . .

Thus, in the community, as in our personal lives, one kind of concern may be hidden in the language of another. This deceit is one form of "lying honestly." The community itself may not know the hidden source or sources of its positions and utterances, whether negative or positive. The transmission of their concerns through local officials adds further distortions. A community or its official body may change its focus or reverse its position several times unexpectedly. For instance, a community may demand assurances of facility safety—a realistic anxiety—but may do so to the point of financial or technical absurdity, demanding a "gold-plated facility," betraying an underlying neurotic anxiety. When someone stands up and demands to know, "How are you going to protect this dump from the Shiite terrorists!" one should not assume that he is stupid. Rather one can assume that risk is not really his interest and that other, more salient concerns (moral, realistic, neurotic) are vainly seeking expression. As Freud (1926, 165–6) so clearly stated,

> In some cases the characteristics of objective anxiety and neurotic anxiety are mingled. The danger is known and objective, but the anxiety in regard to it is overgreat, greater than seems proper. It is this surplus of anxiety which betrays the presence of a neurotic element.

Symbolization of Issues

Many such latent concerns do not have their origin in the siting issue, but are imported, either from within or beyond the community, and projected into it. When a complex of such issues, especially those indigenous to the community, are displaced onto the siting process, the process may be turned into crisis and precipitate hysteria. For example, if a community is under stress due to high unemployment, falling property values, concern over crime, inflation, or concern over the environment, then waste siting, particularly if promoted by a distant governmental body, becomes a symbol. It becomes an object which carries a heavy load of multiple displacements, meanings, and emotions. Free-floating hostilities are displaced upon it. When this occurs a siting project then becomes like the goat in the book of Leviticus (3:12) onto which the whole tribe psychologically projected all its frustrations and sins and then led to ritual slaughter.

Also, those who adamantly oppose a project may intentionally make extreme demands on one level to exploit unspoken fears on another, in effect, lying dishonestly. When this occurs, issues are bled of their substance, actions become ritual, dialogue becomes monologue, monologue becomes soliloquy, soliloquy becomes fantasy. Although one cannot debate a dream-symbol, one can understand it symbolically and learn from it, interpreting it as if it were hieroglyphics (Freud, 1900). A patient explication of its multiple meanings and the motives projected into it may reduce communal neurosis, just as it may personal neurosis (Freud, 1926).

One example of this is the exploitation of the waste issue by environmentalists and other opponents of nuclear power. To a number of such individuals nuclear power is a symbol of evil. Even apart from its easy association with nuclear weaponry and the aura of mystery or fear associated with radioactivity (von Winterfeldt and Edwards, 1984), nuclear power represents centralization, secrecy, complex technology, elitism, techno-rational tunnel vision, life-threatening risks involuntarily imposed, potential destruction of the environment, poor planning, bureaucratic mismanagement, fiscal irresponsibility, socialization of costs and capitalization of profits, and so on. No single attitude correlates as highly with whether one favors or opposes nuclear power as the answer to the question, "Do you think life in this society is just?" Hence, a complex nexus of life injustices (moral issues), logically or not, have affixed themselves symbolically to the cooling towers of nuclear power plants. Whatever drives up the costs of nuclear power, such as the cost of waste disposal, helps drive it off the market. It is no secret that one method of attacking the nuclear power industry is to constipate its waste streams. If this motive is masked in other verbiage in a waste siting project, community emotions are prone to manipulation and dialogue remains inauthentic.

Failure to provide forums for the expression of irrational emotions, thoughts, and values, including displacements and symbols, and the failure to listen to them carefully, may eventually come back to haunt a siting process. Many siting agents, habituated by the nature of their work to the sterile language and narrow technocratic logic of science and bureaucracy, turn their minds off when confronted with the seemingly irrational, saying to themselves, "Go ahead, fool, vent. When you are done screaming and shouting, *then* I'll listen." Or they say, "That's a political issue" or "That's a value judgment," as if political issues and values are incapable of understanding or rational discourse. Many technocrats are unable to make an effective response to irrational statements because they do not have the training which would enable them to do so.

Because neurotic anxiety motivates both individual or community actions, the seemingly irrational act is not where understanding stops, but where it is most essential. If these preconscious concerns are expressed openly, understood, and addressed, as in the psychoanalysis of neurotic patients, then a siting process might avoid much waste of time and money, fear, misunderstanding, and failure. On the other hand, if they are merely taken at face value, ignored, or left obfuscated, the prospects for meaningful, fruitful public participation and cooperation are lessened. Failing in one community and moving to another, only to repeat the same approach, is mindless bureaucratic ritual at its worst.

Psychoanalytic Vs. Strict Empiricist Theory

Bureaucracy worships technocracy with its devotion to strict empiricism and its promise of control and order as its *Weltanshauung,* its religion (Weber, 1947). Hence, traditional approaches have shunned rather than invited encounters with the irrational. The irrational threatens the bureaucrat's worldview of order, prediction, and control. Examination of the siting literature and attendance at conferences and workshops on siting and community incentives reveal that neurotic anxieties and their psychosexual origins are a wholly neglected force. One reason is that traditional theoretical approaches and conceptual nets define "reality" and "reason" in narrow, scientific terms and are unable to capture the complexity of subtle forces at work in a community waste siting process.

A systematic contrast of the psychoanalytic theoretical approach with all the others is not possible here. Major approaches include: (1) the traditional issues approach (Safety, Equity, Trust, noted above); (2) the institutional mismatch approach which seeks to shift the task to a different level of government; (3) the cognitive dissonance theory approach which seeks to reconcile public risk perception with scientific risk assessment; (4) the functional theory approach which relies on public information, systems logic, and appeals to the common good; and (5) the behavioral theory approach which seeks to entice hosts with sufficient incentives and rewards. In practice, siting efforts are usually an amalgamation of several of these (Lake, 1987). The nuances in ontological and epistemological assumptions would constitute a separate book. Here a few paragraphs will have to suffice.

These approaches share in common the traditional ontological and epistemological assumptions of science—value-freedom and objectivity. These are inappropriate or insufficient when applied to human beings. Humans are neither value free nor objective. In addition to ideas (ego), individuals and communities are bearers of and driven by both passions (id) and cultural values (superego). (See Fig. 12-1.) Thus, the techno-rational perspective blinkers out essential realities. For example, the implicit assumption that human beings are only or even primarily motivated by conscious ideas, much less by techno-rational ideas, creates a limited, one-dimensional model of man, one which by default is mechanistic and demeaning (Marcuse, 1964). Another bias in the techno-rational perspective is the implicit assumption that basic human emotions are somehow less stable than ideas, rules, laws, etc. On the contrary, all of these can change over night and, in fact, do when conditions merit (Fox, 1967). In contrast, psychoanalytic theory recognizes that it is often easier to change people's ideas or values than it is to change the basic emotions or fixations which drive them. (Cf. smoking literature.)

A practical consequence of techno-rational assumptions when applied to waste siting, is that siting agents, most of whom have technical backgrounds, believe that if people are presented with enough empirical facts—a large enough dose of reality principle—then the natives of a local community will see things their way. Not so. Public information does not necessarily lead to public persuasion. For example, there is no correlation between citizens' knowledge of radioactivity and their attitude for or against nuclear power (Hensler and Hensler, 1979). Facts do not necessarily change values; education is not conversion. Facts never speak for themselves; anxieties do.

One perspective, currently in vogue, nicely epitomizes the limitation to understanding which techno-rational epistemology entails—the behavioral approach. This popular approach attempts to overcome community resistance, not by addressing and resolving issues, but by offering irresistibly attractive financial incentives to overlook them. Some would pretend that this buy-a-site approach is novel and now repackage it as if it were innovative: "... we propose a fundamentally new approach to facility siting—negotiated compensation for local disamenity..." (O'Hare et al., 1983, vii). Promises of jobs, scholarships, new roads, guaranteed property values, tax revenues, and so on, are offered in exchange for local acquiescence. Hope is often expressed, though seldom publicly, that the search for a site should focus on a rural county which is economically depressed, the assumption being that people in economically hard pressed areas will more readily succumb to the temptations of such financial incentives. Further, being poor and sparsely populated, such counties are less able to mount effective political resistance. Often counties where hazardous waste facilities are located are populated predominately by Blacks (USGAO, 1983).

A strictly behavioral approach eschews any concern with a community's emotions, ideas, or values, defines itself as being beyond issues of human freedom and dignity, and therefore seeks merely to alter behavior toward the ends desired by those in power (Skinner, 1971). Waste siting "success" rests solely on whether a site is obtained, without regard to individual psychological harm or the sociological derangements inflicted on a community in the process. There are two powerful reasons why bureaucrats embrace the behavioral approach: (1) it works and (2) it is empirical and objective. It avoids the subjective realms of politics, emotions, beliefs, and values. It also has the added advantage that, having thus enticed a volunteer to host a site, the Equity Issue appears resolved *quid pro quo:* "You may put it in my backyard for a price."

In the past, the buy-a-site approach often succeeded. In recent years, however, it has succeeded less and less. Thirty-two different hazardous waste siting efforts were made nationally between 1979 and 1984. Of these, five treatment and two storage facilities were approved, but only one waste disposal facility was built (Maryland) and that is now closed (Kasperson, 1985). In addition to realistic anxieties triggered by technical failures at older sites, the moral issue of whether it is appropriate to exploit the disadvantaged has increasingly come to the forefront (USGAO, 1983). For example, in an economically depressed region of south Texas, a contract for construction of a site for the disposal of low-level radioactive waste was near consummation in 1985. The Catholic Church at the last minute rallied the poor against the project and the site was abandoned (Blackburn, 1985).

The court-ordered multimillion dollar exhumation of hazardous wastes in rural Wilsonville, Illinois, was referred to earlier. The site was approved and permitted by the Illinois EPA and was in full operation with no violations of discharge criteria. The presence of jobs and tax revenues provided by the site did not offset a violent community backlash. When the nature of the wastes being buried became known to the locals, only a court injunction prevented violent insurrection.

> By Sunday evening [in April, 1977], when an angry armed mob gathered to stop the trucks importing the PCB-contaminated soil, local leaders had induced a retired circuit court judge to act as the village's special attorney to initiate legal action to stop the PCB importation on the condition that there be no violence. The mob dispersed, and the following day Wilsonville began legal action (O'Hare et al., 1983, 20–21; USEPA, 1979a, 307).

The armed mob was comprised largely of Wilsonville residents and area farmers with shotguns and hunting rifles. According to a local priest, some carried sticks of dynamite. These examples suggest that the neglect of the underlying moral, realistic, and neurotic anxieties implicit in the behavioral approach and its theoretical cousins, may be self-defeating to waste siting projects.

The techno-rational theoretical paradigm, one which appears "value-neutral," "scientific," "empirical," and "objective," is philosophically compatible with the *Weltanshauung* of government and corporate bureaucrats. Contractors and consultants, adept at dense techno-rational language, make fortunes selling government bureaucrats what they want to hear, despite the fact their programs or products are often failures. In siting projects, a fortune is spent in obsessively preparing color slides, charts, graphs, reports, studies, models, and risk assessments, while little if any effort is spent in preparing for a discussion of community emotions or fundamental values. These fall outside the bounds of techno-rational reality.

Ironically, the problem of waste disposal is not a technical problem. The technical knowledge necessary to build disposal facilities of all types has advanced rapidly and is demonstrated in many countries. Rather, waste disposal is a *social* problem and requires an anthropological, humanist perspective, one which incorporates community values and emotions. Unfortunately, social scientists who are attracted to the issue of waste siting often share the same quantitative, techno-rational epistemology as the technically-oriented bureaucrats who hire them (Otway and Thomas, 1982). Such an echo environment of narrow or false assumptions leads to a tissue of institutional self-delusions which eventually collapses. We witness the symptoms in the confessional sequence of organizations: lying honestly ("Our policy is morally opposed to such a course"), then loss of memory ("It never happened"), then plausible deniability ("It happened but we didn't do it"), then damage control ("We didn't know. A few bad apples may have been responsible"), then taking the fifth ("No comment") then "copping a plea" ("Guilty, your honor"). Wrote Freud (1921, 80),

> [G]roups have never thirsted after truth. They demand illusions, and cannot do without them. They constantly give what is unreal precedence over what is real; they are almost as strongly influenced by what is untrue as by what is true. They have an evident tendency not to distinguish between the two.
>
> If we consider mankind as a whole, ... we discover that it too has developed delusions which are inaccessible to logical criticism and which contradict reality. (Freud, 1937, 269)

THE ANAL STAGE OF DEVELOPMENT

The answer to the question, What makes waste different?, is simple: It is waste. From the study of infants and small children and from cross-cultural studies by anthropologists there is no question but that the repulsion which, as adults, we feel toward defecation and feces is due primarily to the socialization process. Small children and infants show no innate aversion to their own excreta. Indeed, quite the opposite. They delight in playing with it (Freud, 1930, 100n). In polite society one is not to speak of such things,

> Nevertheless, psychoanalysis and folklore have not allowed themselves to be deterred from transgressing these prohibitions and have been able as a result to teach us all kinds of things that are indispensable for an understanding of human nature.... In the earliest years of infancy there is as yet no trace of shame about the excretory functions or of disgust at excreta. Small children show great interest in these, just as they do in others of their bodily secretions; they like occupying them-

> selves with them and can derive many kinds of pleasure from doing so. Excreta, regarded as parts of a child's own body and as products of his own organism, have a share in the . . . narcissistic esteem . . . with which he regards everything relating to himself. Children are proud, as it were, of their own excretions and make use of them to help in asserting themselves against adults (Freud, 1913, 336).

Studies of primitive peoples also reveal a wide variation in treatment of the topic. Cross-culturally, coprophilic interests are found in magical ritual, in tribal customs, in religious observances, and in the art of healing (Bourke, 1891). For example, in other cultures feces are used for decoration of the face by members of the bridal party at weddings as a matter of custom. In sum, psychological aversion to excreta is overwhelmingly man-made, not natural. It is a culturally-induced abhorrence typical of civilizations.[4]

Thus, the socialization process is confronted with the precarious task of transforming a neutral or positive predisposition into a wholly negative one. This makes the repression of anal eroticism during the anal stage one of the more demanding and delicate in terms of importance to the ego of any erotogenic stage. The possibility of unbalanced or unsuccessful repression of anal eroticism is therefore strong. Not only are fixations (Freud, 1916–1917, 340; Rycroft, 1968, 52) likely to occur, but, because a complete reversal in attitude is sought, all traits connected with waste are etiologically *highly ambiguous* and *unresolved.* Because such positive desires are not eliminated but merely repressed into the unconscious, an ambivalent attitude toward waste generates strong reaction-formations (Horowitz, 1977, 40). Of the two classic types of fixation, retentive or expulsive tendencies, we are concerned here only with the reaction formations associated with the retentive tendency.

The Bedrock of Psychosexual Resistance

In our aversion to waste or defecation we do not have the luxury of a one-sided hatred such as that which many people develop for snakes. It

[4]Freud (1930, 100n) does suggest, in part, an organic basis, but one shrouded in the mists of evolutionary history: "Such a reversal of values would be almost impossible to bring about, were it not that these substances expelled from the body are destined by their strong odours to share the fate that overtook olfactory stimuli after man had erected himself from the ground. Anal erotism, therefore, is from the first subjected to the 'organic repression' which paved the way to civilization [via upright posture and year-round mating]."

is easy to hate that which one seldom has to deal with and which, when one does, offers no pleasures. But each of us deals daily with his own wastes and their elimination from the body and, furthermore, it is often a pleasurable experience. Far from causing a phobic reaction, we enjoy it. Only our repressions keep us from admitting it publicly. We are taught to be ashamed of the act, to be silent about it, to conduct it in utmost privacy, and not to discuss it (Freud, 1913, 336).

This ambivalence toward defecation is overlaid with early childhood repressions associated with the product itself, the excreta. We do not like to be reminded of our repulsive attitudes toward feces. The reason stems from the harsh and painful injunctions, including a sense of dirt and shame, inflicted upon us by our parents, persons upon whom we were totally dependent and upon whom we later came to bestow awe, love, and affection (Freud, 1923, 36). Our repugnancy of waste as something dirty and forbidden and our shame in associating ourselves with it, is little else than the residue (unconscious memory traces) of that ingrained pain, about which as adults we now enjoy total amnesia (Freud, 1905, 174–6). Unconsciously we do not want to recall or be reminded of those negative experiences.

With respect to both defecation and excreta, attitudes which are ambivalent and tenuous in their origins become equally strong in their reaction formations of shame, disgust, and moral outrage (Freud, 1908, 171). Here we find the bedrock of community resistance to hosting a hazardous waste site. The resistance is a defense against *the return of the repressed* (Freud, 1915, 154; Freud, 1939, 124 ff.; Horowitz, 1977, 34); that is, a defense against any involuntary irruption into consciousness of unacceptable primary impulses, such as anal eroticism, and their derivatives (Rycroft, 1968, 142). Psychologically the thought of the return of this repressed object unconsciously and simultaneously represents both painful experiences and forbidden wishes.

As shown in Figure 12-2, the thought of "the return of the repressed" material triggers

- (id-level): unconscious horror which causes conscious revulsion which may be acted out as hysteria or anarchy;
- (ego-level): unconscious fear which causes conscious anger which may be acted out as violence, sabotage, or threats of violence; and
- (superego-level): unconscious moral disgust which causes conscious outrage and righteous indignation which may be acted out in legal resistance or civil disobedience.

PSYCHOANALYTIC LEVEL	REACTIONS: Unconscious	Conscious	Action
SUPEREGO:	DISGUST	RIGHTEOUS INDIGNATION	LEGAL RESISTANCE OR CIVIL DISOBEDIENCE
EGO:	FEAR	ANGER	VIOLENCE AND THREATS OF VIOLENCE
ID:	HORROR	REVULSION	HYSTERIA AND ANARCHY

Figure 12-2. Affective reaction formations to the return of the repressed.

The Soft Voice of Reason vs. A Compulsion to Repeat

An objection may be raised that feces and industrial wastes are by no means the same; that, in fact, in terms of disposal they are entirely different issues; that any connotative association between them, however powerful it may seem to the unconscious mind, is clearly false to the rational one; that assurances of the benign nature of the waste or, at least, of its secure technical and institutional containment can be impressed upon a wary citizenry; that, in fact, no danger-situation exists; and that, therefore, rational persuasion, intelligence, and patient education can undo any serious mental or emotional errors which the citizens of a community may make. To this objection may be added another strong argument, namely, the empirical fact that, whatever the danger a controlled facility really poses to health and safety, it is certainly far less risky than other dangers voluntarily assumed by citizens and communities every day.

On this rock waste management builds its church, while projecting its sins onto other scapegoats. One hears it expressed, "There is really nothing wrong with these people if you give them a chance to understand. The problem is really only a few hotheads, some self-serving politicians trying to capitalize on the issue, a few well-meaning but misguided environmental types looking for a cause, and, of course, those irresponsible sensationalists in the media who scare everybody and don't care to get their facts right."

Assume for the moment that the risk assessors are correct and that citizens' sense of danger is largely delusional. If so, waste siting, at the collective level, is analogous to the problem faced by the psychoanalyst in aiding his deluded patients to surrender their phobias and compulsions. What do they tell us? Despite the fact objective reality is on the side of the sane, psychoanalytic practice indicates that the task is arduous and prospects of success are often grim. This is because

> ... the nature of repression ... is fundamentally an attempt at flight. What is repressed is, as it were, outlawed; it is excluded from ... the ego and is only subject to the laws which govern the realm of the unconscious. If, now, the danger-situation changes [such as via improved waste technology] so that the ego [community] has no reason for warding off a new ... impulse [that is, accepting a new waste facility] which resembles the repressed one [either a bad experience with the old waste dump or, more deeply, its repressed collective coprophilia], the consequence of the restriction of the ego which has taken place [a community's or person's waste aversion] will become manifest. The new impulse will run its course under an automatic influence— ... repetition-compulsion. It will follow the same path as the earlier, repressed impulse, as though the danger-situation that had been overcome still existed.
>
> The ego [or community] may occasionally manage to break down the barriers of repression which it has itself put up and to recover its influence over the instinctual impulse and direct its course in accordance with the changed danger-situation. But in point of fact the ego very seldom succeeds in doing this: it cannot undo its repressions.... That this is the origin of ... the retention of danger-situations which are no longer present-day ones is confirmed by the fact of psychoanalytic therapy (Freud, 1926, 153; bracketed comments added).

It is confirmed by the general failure of waste siting projects as well. To the extent industrial or radioactive waste is connotatively linked, rightly or wrongly, to feces or to older waste sites which have failed, one should expect the soft voice of reason to be overruled by the unconscious law of repetition-compulsion fears. The church is built on sinking sand rather than rock.

Finally, just as the child fears reproach and loss of love when he soils, the same fear occurs in a potential host community, heard in this excerpt from a typical protest letter to the editor:

> I personally don't think we need the dozen or so jobs it (waste disposal site) will bring so badly as to scare off others. Many families would move out of the area. Others thinking of locating here would instead go

> elsewhere. Some businesses would leave. Others would pass us by. Some industries would no longer look us over (Anonymous, 1987).

For the community, this expresses in polite terms the underlying psychological fear which cannot be publicly expressed: "A person who is not clean—who does not hide his excreta—is offending other people; he is showing no consideration for them. And this is confirmed by our strongest and commonest terms of abuse" (Freud, 1930, 100n).

Infantile Consideration for the Object

During the anal stage, in order to secure the parents' love, the child must learn to control its sphincters, to renounce certain anal pleasures, and to give as well as to receive. Overlapping this period is the development of the child's "consideration for the object" (Freud, 1933, 64). The ambivalence experienced toward defecation and excreta described above, is carried over to the child's consideration of objects, including persons, encountered in the outside world. The child's relationship to external objects occurs in two stages. The first is characterized by partial incorporation of the object; the second, by domination of the object. Both stages are marked by ambivalence.

First, recognition of the separate existence of objects advances to a point where the object or person is represented by only a portion of itself, and it is this part, rather than the totality of the object, that is in fantasy incorporated into the self. This may, for instance, take the form of trying to suck in, recycle, or eat the object. The child regards the object

> . . . as something over which he exercises ownership and that he consequently treats . . . in the same way as he does his earliest piece of private property, i.e. . . . his faeces. . . . And since the ambivalence of feelings still exists in full force . . . he expresses his positive attitude towards his object in the form of retaining his property and his negative attitude in the form of rejecting (expelling) it (Abraham, 1924, 426).

Expulsion of the part object represents the removal or loss of an object, thus its destruction. The positive tendency to preserve or retain the object steadily grows in strength until a dividing line is reached where ". . . the tendency to preserve the object begins to predominate" (Abraham, 1924, 432). By the end of the first stage, "The object is now consistently spared from destruction, though it remains the target of ambivalent feelings" (Horowitz, 1977, 60).

Second, the stage of partial incorporation is replaced by a desire for control over the whole object. Instead of taking part of the object into the self and interiorizing it, the child wants to rule and possess it externally. The object is still possessed, but now as an exteriorized object, one recognized as existing outside the self, though still to be ruled, controlled, and possessed. Here adulthood possessiveness, domination, and control over things and people as functional objects have their embryo (Finney, 1961; Adorno, 1950.)

The inevitable frustration and hostility which the child experiences at the incorporative stage are not resolved, but merely changed, by advancing to the domination stage (Horowitz, 1977, 61). This second stage is also contradictory, making the anal stage fully ambivalent throughout. The desire to dominate the object or person contradicts the recognition of the existence of the object or person as a separate entity. Thus, individuals fixated at the anal stage manifest reaction formations against this ambivalence toward waste as they move from a period of incorporation (recycling) to one of domination, i.e., a refusal to let go.

The history of the management of U.S. commercial high level radioactive waste reveals a remarkable parallel. First, the nuclear industry gave its attention to part object, that component of the fuel rod which can be reused, and attempted to incorporate it back into the system by recycling (reprocessing). When the whole object came into view, it was apparent that reprocessed spent fuel rods create a highly dangerous material, near weapon-grade plutonium. For national security reasons, Presidents Ford and Carter outlawed reprocessing and dashed what was left of the fantasy of a "plutonium economy."

Second, this desire for partial incorporation graduated to a "tendency to preserve the object," marked by a desire for control and domination. President Reagan lifted the ban on reprocessing yet, unlike other countries, we neither bury these wastes nor reprocess them. At this stage, "The object [spent fuel] is now consistently spared from destruction [either by reprocessing or disposal], though it remains the target of ambivalent feelings."

An Example of Reaction Formation in the Waste Industry

Given Freudian theory, one should expect to find a reaction formation by those in waste management against the one publicly acceptable word in English most associated with both defecation and industrial waste disposal: dump. In the waste management business the avoidance of the word "dump" as if it were a cultural tabu betrays its psychosexual connection. Never does one publicly find the director of a waste siting project involved on a dump project. In techno-rational papers, reports, and proposals written by consultants, professors, and researchers for state and federal agencies, never is a dump a dump. It is a "sanitary landfill" or a "low-level radioactive waste disposal facility" or "a spent fuel repository project."

The flight from the neurotic anxiety inherent in the topic of waste elimination by those in waste management sometimes borders on the comic. A manager in one organization responsible for a waste siting project instructed his staff to search out a substitute word for "dump" in hopes the editors of local newspapers would not use it in their articles and headlines. He complained that newspaper editors use "dump" because it is a four-letter word, meaning it is short and fits easily in a headline. He seemed oblivious to his own Freudian slip, his wish that "dump" were one of those four-letter words we cannot publicly speak. In effect, the newspapers were dumping on his project: "The editors use 'dump' because it is a four-letter word."

Trivial in itself, this transparent reaction formation, a word tabu intended to suppress a psychosexual connection, reveals more than it hides. Namely, waste managers seek to define reality by their wishes so that they can lie to themselves and others "honestly" about the true nature of their work. In so doing, they are as irrational and self-serving as the public and environmentalists whose "irrationality" they so robustly criticize behind closed doors. In a waste siting process the "palpable device" of using technically sterile, sanitized language often heightens a community's moral and realistic anxieties, particularly with respect to the Trust Issue. "What are they hiding with all that fancy, technical talk?" thoughtful citizens ask. These, in turn, trigger other anxieties which boomerang as community resistance. A member of one waste siting staff cynically asserted in private that an important criterion for his job was the ability to lie publicly without blushing. But from within the bureaucracy, there reigns a conspiracy of silence: "Yet who so bold but says he sees it not!"

Honesty, however, sometimes leaks out, if only through other metaphors. There is, for example, the refreshingly frank reference in the industrial literature to waste as a problem at the "back end" of the process such as the nuclear fuel cycle. Indeed, for each of us it was at the back end of our fuel cycle where our waste management first began. Among other traits, individuals with anal retentive fixations show marked "interests in the backs of things" (Kline, 1972, 10).

This tabu is also instructive in that it occurs exclusively in the waste management subculture. Newspapers, environmentalists, or the public at large do not find the word "dump" publicly offensive and do not hesitate to use it. It is an hypothesis that those who are attracted to waste management as careers can be shown to have over-determined anal reaction formation characteristics, not the least of which is a hatred of waste (Kline, 1972, 11). One would predict both higher than average somatic illnesses associated with the bowel in this group, a stronger aversion to homosexuality, and a stronger resistance to the psychosexual connections set forth in this paper (Freud, 1925, 217–222).

In sum, much as the garbage man prefers to be called a "sanitation engineer," it is an irony that those in government and industry who have it as their mission to eliminate civilization's industrial filth, sanitize the idea. This understandable, yet self-defeating attitude towards the reality of the task is traceable to the ambiguous etiology of waste neurosis. In these examples also lies hidden the deeper, counter-intuitive thread that "... cleanliness is far more readily associated with vice than with virtue ..." (Freud, 1913). The trait of obsessional neurotics to compulsively clean is empirically traceable to anal character traits (Rosenwald, 1972; Blatt, 1964; Schlesinger, 1963; Segal, 1961; Sandler and Hazari, 1960). To these deeper issues, we now turn.

Waste Bias

Many will recall from their teenage years that there was hardly a harsher condemnation which one could say about someone else than the colloquial expression, "He thinks his shit don't stink." As it turns out, he probably doesn't. Each person, noted Freud (1930, 100n), "... scarcely finds the smell of *his own* excreta repulsive, but only that of other people's." This is certainly true of our other body wastes, such as sweat, body odor, and breath odor, and there is no reason it would not be true with respect to excreta. Thus, the person in the room who stinks the most may indeed be the last to know, a fear frequently exploited by the marketeers of hygienes, particularly on television.

Seeing someone else's neglected feces floating in a toilet in a public restroom is immediately repulsive. Many find this so disgusting that they refuse to even use that toilet. But when they themselves have performed their function and see their own product, no such aversion occurs. This peculiar waste bias in favor of ourselves and against others, suggests that each person in a community has an unconscious neurotic anxiety with respect to other people's (society's) waste, far more so than if, in some sense, he could identify it as his own.

An illustration. Illinois hosts the largest commercial nuclear utility in the United States. Commonwealth Edison has twelve nuclear reacors in operation at six separate sites, all in the northern third of the state. Citizens express remarkably little concern over the fact that each nuclear power plant must stockpile its highly dangerous spent fuel rods on-site. However, if Commonwealth Edison attempts to transfer these spent fuel rods from one nuclear plant to another, a distance in some cases less than 100 miles, local communities react vociferously. So strong is this bias against the importation of "outsider's" waste that each nuclear plant in Illinois must provide its own spent fuel storage pool. Thus, plant-to-plant shipments are not done. At the same time, shipments of identical radioactive wastes, to and from nuclear facilities in other states, cross Illinois as a matter of unremarked routine. No amount of risk analysis or other empirical approaches can explain this willingness to strain at the gnat of local shipments while swallowing the two-humped camel of local on-site storage and interstate shipment. Freudian psychology, not risk analysis, provides the explanation.

This unconscious psychological bias against other people's waste aids in understanding both a community's strong resistance to hosting a regional or national waste site and the difficulty in getting a community or its officials to "own the problem." The Equity Issue—"Take it back where they made it"—is, in part, a manifestation of this deeper psychosexual aversion. Colloquially, the anal connection is clear. "Take it back where it came from," says the local citizen in anger. "We are," says the sociologist. "It was made right here in the United States. And this is where we want to put it. Aren't you proud to be a part of your country?" The angry man answers, "Don't give me that crap. You take it and stick it where the sun never shines." Answers the sociologist, "That's exactly where we want to put it—as you suggest, in the dirt—buried, where the sun never shines and the wind never blows." Shots are fired. "*There's* one less wise-ass," says the citizen with smug satisfaction. Soon after, local bumper-stickers, intended to offend, sneer out with mock fatalism, "Shit Happens."

The above vignette, in raw language, illustrates the thesis of the overall argument:

> . . . the history of the first prohibition which a child comes across—the prohibition against getting pleasure from anal activity and its products—has a decisive effect on his whole development. This must be the first occasion on which the infant has a glimpse of an environment hostile to his instinctual impulses, on which he learns to separate his own entity from this alien one and on which he carries out the first "repression" of his possibilities for pleasure. From that time on, what is "anal" remains the symbol of everything that is to be repudiated and excluded from life (Freud, 1905, footnote added in 1920, 187n).

Reality Principle vs. Pleasure Principle

Culturally, the gulf between what we want and what we must do is widening. The potential for explosive conflict increases proportionally. Experience nationwide suggests that someone in the U.S. may be harmed or killed in conflict over waste siting. As Kasperson notes,

> The vehemence of the opposition often stuns facility developers and regulators, as in one example where angry citizens were prepared to blow up a facility and there were reports of threats of death or physical harm to key individuals and their families (USEPA, 1979, iii).

The range of waste management options in both time and space is limited and shrinking rapidly. Regardless of the kind of waste, disposal space is at a premium nationwide. This, despite the fact public health and safety—the reality principle—demand intelligent management and disposal.

This lethal disjuncture is part of a larger cultural process. As Freud (1925, 219) pointed out,

> Society maintains a condition of *cultural hypocrisy* which is . . . accompanied by a sense of insecurity and a necessity for guarding what is an undeniably precarious situation by forbidding criticism and discussion.

A Freudian analysis of the cultural process by which *The Culture of Narcissism* (Lasch, 1978) led to *The Closing of the American Mind* (Bloom, 1987), under the tutelage of *Friendly Fascism* (Gross, 1980), is beyond the intent of this paper, except that its consequences to waste siting, such as the public's loss of faith in basic institutions, are germane. In recent decades the pleasure principle has dramatically and progressively encroached upon the reality principle, as evident in increasing ethnocentrism, provincialism, narrowed conceptual horizons, a narcotic of pseudo-happiness, and an upsurge in creeds of self-interest, both religious and secular. Metaphorically, a culture which sings "Don't worry, be happy," mutes serious criticism and discussion. As a society, ego is increasingly in bed with id, making culture less and less ideational and increasingly sensate (Sorokin, 1947). This cultural hypocrisy has exacerbated attempts to realistically address a variety of social problems, particularly waste management and disposal, "an undeniably precarious situation."

Waste management is not a topic that attracts the passions of charismatic or innovative leadership. No Master's degrees or Ph.D.'s are offered in the subject. This benign neglect results in short-range political expediencies. In 1980, Congress, under political duress from three governors, devised a system of regional compacts for the disposal of low-level radioactive waste. Instead of one cost effective site for the waste stream of the entire nation, this plan puts one site in Massachusetts, New York, Pennsylvania, North Carolina, Illinois, Michigan, Nebraska, California, Texas, and so on, at enormous expense and multiples of social strife. In a second cycle, additional facilities may be located in Virginia, Ohio, and elsewhere. If this balkanization of the waste stream succeeds, the nation can expect to maintain between 12 and 20 radioactive waste sites, almost all of which will be in humid climates and major watersheds, for the next 300 years. Such ill-advised approaches often flounder for a few years, until their political or technical unworkability becomes apparent. They then are politically orphaned.

The tendency in a market-driven economy to prefer the pleasure principle, such as cheap electricity and large profits, to the reality principle, such as recognition of the costs associated with providing for the permanent burial of spent nuclear fuel and other radioactive wastes, sets up a structural tension which can result in organizational paralysis. The management history of commercial nuclear spent fuel is illustrative. In the last 30 years, the U.S. nuclear power industry found the motivation, political will, technical know-how, and wherewithal to construct more than 110 commercial reactors, but not one waste site for spent fuel. All that time, the industry has been "holding in" its commercially generated spent fuel rods. These rods are consolidated, provided new pools, racked, and reracked. Every five years an alarm sounds that the nuclear industry is running out of spent fuel storage space and that a permanent solution to the disposal problem must be found, after which nothing is done other than to add another five-year capacity to existing pools. The commercial nuclear power industry gladly transfers responsibility for the disposal of this waste to the federal government, while passing through its costs to the ratepayers. What have we to show for the hundreds of millions of dollars expended? Nothing. Not one rod is in the ground.

The psychoanalytic perspective finds the same cause on both sides. Community resistance to hosting a site has the same origin as the resistance by industry or government to finding one—ambivalence manifest as political constipation. In our culture, as in our minds, the issue of waste remains ambiguous and unresolved. We choose, if possible, not to think about it, not to deal with it. Again, the description which Horowitz (1977, 60) gives of the child's anal stage attitude toward objects can be applied to the present relationship with spent nuclear fuel and other types of waste: "The object is now consistently spared from destruction, though it remains the target of ambivalent feelings."

Summary of Industry vs. Community

With respect to industry and governmental agencies responsible for waste disposal (elimination, destruction), one finds reaction formations in the form of denials, a long-standing ambivalence toward the object, a sanitization of language, and a costly extended "institutional constipation." Waste disposal programs are marked by a pandering to provincialism, expediency of design, theoretical blinkers, *de facto* class or race exploitation, and an ambivalence of will which is often wholly incapable of overcoming local resistance.

From the community perspective, there are strong reasons on all three levels—id, ego, superego—why deep, unconscious anxieties are triggered by the suggestion that one should *take in* waste rather than expel it. Especially someone else's. The strong resistance encountered in waste siting in the U.S. is a cultural phenomenon that expresses this complex mix of moral, realistic, and neurotic anxiety. From a Freudian view, in terms of our collective psychosexual development, community political resistance, anger, hostility, and hysteria are predictable reactions to the return of the repressed. Such reactions by a community to the return of the repressed are merely the collective analog of the personal resistance observed in patients undergoing psychotherapy, whence the term originally sprang (Freud, 1915; cf. Jung, 1918; Fenichel, 1945; MacKinnon and Dukes, 1962). Freud (1896, 170) first coined the phrase—appropriately enough—in his remarks on the neuropsychoses of defense.

ANAL CHARACTER

A distressing duty which very young children must endure as a part of toilet training is holding in or retaining their excreta until they are permitted or instructed to eliminate it. Failure to do so becomes an occasion for social shame and upset, especially at school. This command to "hold in" or retain one's excreta until allowed to do otherwise may not only be unpleasant and physically difficult to master, but also can result in a fixation upon anal retention. This is particularly true if frustration and renunciation are imposed on a child prematurely, suddenly, severely, excessively, or inconsistently (Horowitz, 1977, 39).

Freud's hypothesis that "Important constituents in the formation of character are developed, or strengthened, from the repression of coprophilic desires" (Freud, 1913, 337) has found uneven empirical support. This is due, in large part, to methodological problems inherent in nonclinical research (Kline, 1972; Fisher and Greenberg, 1977). What has found strong and repeated empirical support is the cluster of traits which Freud first identified as anal character (Freud, 1908, 169–175). From these fixations and the traits associated with them via displacement and reaction formation, we can identify a composite of characteristics expressed both in personality and in society.

Freud (1908, 169) first described the anal triad as follows:

> The persons whom I am about to describe are remarkable for a regular combination of the three following characteristics. They are exceptionally *orderly, parsimonious,* and *obstinate.* Each of these words really covers a small group or series of traits which are related to one another. "Orderly" comprises both bodily cleanliness and reliability and conscientiousness in the performance of petty duties: the opposite of it would be "untidy" and "negligent." "Parsimony" may be exaggerated up to the point of avarice; and obstinacy may amount to defiance, with which irascibility and vindictiveness may easily be associated.

Just as the oral-fixated individual displaces his interests onto cigarettes or food (Jacobs et al., 1965, 1966, 1971), the anal-fixated displaces his retentive interest onto those things which one can hold in, that is, hoard—collectables, especially solid objects. One may observe in the anal retentive character a compulsion to collect books, stamps, or especially money. A compulsion to save money and save time (Campos, 1966; Pettit, 1969) are obvious variants of the parsimony attributable to personalities with a retaining tendency. To the anal retentive, to waste time or waste money is more than a loss. It is a shame.

The connection of money to feces, or money as feces displacement, finds expression in custom, myth, colloquial expression, psychotherapy, as well as support in empirical research (Schlesinger, 1963; Noblin, 1962; Lerner, 1961; Rosenwald, 1972). Those who hoard money and are stingy with it are colloquially called a "tight ass" or a "tight wad." In legends from both China and ancient Babylon, gold is referred to as "the excrement of the Devil" (or underworld). The Bible associates money with filth, which perhaps explains why in Christmas gift-giving one carefully removes all price tags as if hiding a moral stain. To "do one's business" in toilet training is more than metaphorically related to the essentials of business—making money. As Menninger (1945, 172–3) notes:

> For many individuals the most important sublimation of this period is the "making" of money. It is probable that one's attitude towards money is formulated by events in the anal stage but the earning of it is definitely related to this phase. Probably no other material product is so closely and universally equated to feces as is money. This association is indicated in the fairy story of the goose that laid the golden egg. [To "goose" someone means to suddenly stick your finger in his anus.] It is recognized in the common expressions of "rolling in wealth," "stinking rich," "a shitpot full of money," a "pot" of gold, (and) "filthy lucre." (bracketed comment added)

The moral outrage expressed in communities over the offer of money (tax revenues, jobs, scholarships) in exchange for hosting a toxic or radioactive waste site is rooted in this unconscious association of money with waste. Such offers are often considered to be "tainted" or "dirty deals" or outright bribery. "We don't want your money and we don't want your waste!" Yet these same communities host many other hazardous or noxious facilities *quid pro quo,* such as oil refineries, gasoline stations, grain elevators, and coal mines, with little or no such reaction.

Traits of the Anal Character

If the retaining fixation is manifest in full reaction formation, a personality is marked by a preoccupation with orderliness, cleanliness, reliability, a delight in organizing, a pedantic attention to detail or rules, a love of symmetry and exactitude, and, not the least—as we should expect—an abhorrence and hatred of messiness or waste (Kline, 1972). Such individuals often think of themselves as perfectionists and, in obsessive-neurotic form, may compulsively wash their hands or clean things.

The above traits derive from an interest in the product of the act. Expanding upon the three traits first noted by Freud in 1908, researchers have identified additional character traits derived from the act itself (Jones, 1923; Menninger, 1945). These are included in the composite which follows, along with anal character traits as shaped by the two defense mechanisms, displacement and reaction formation.

The anal character tends to procrastinate, then concentrate on a topic or task intensely, which results in an atmosphere of perpetual crisis and manic productivity. He therefore does not like to have his time used up by others. He believes that nobody else can do anything as well as himself and therefore has an inability to depute work to others. His obstinacy is manifest as an intolerant insistence in doing things one way—his way—marked by a boring persistence to see routine tasks done. To "do one's duty" as a young child in toilet-training carries over directly into doing one's duty in adulthood. Where duty is not clear, however, and decisions are called for, he is indecisive. As components of a self-willed independence, a love of self-control, and a highly individualistic style, he resents any advice from or thwarting by others and will oppose any attempts to guide his conduct. These traits make him easily irritated by others. His tendency to domineer is oddly compensated by a special tenderness toward children. He overreacts to any criticism or injustice shown him and, in turn, he will show vindictive desires for revenge against those who thwart or challenge him. At times, he even may become sadistic in his unfeeling treatment of others. Because of his minute attention to detail and a strong drive to see things orderly and clean, his pleasures are spoiled by trivial things such as small objects out of place. He may show an interest in the backs of things. Finally, he defends himself by a strong belief in rights and individual dignity (Kline, 1972; Menninger, 1945; Jones, 1923).

In sum, these are the cluster of traits built up from the repression of anal retentive tendencies (fixations) and manifest in individual character. Almost all can be generalized to groups.

ANAL CULTURE

Freud's (1913a, 190) insight that "Our highest virtues have grown up, as reaction formations and sublimations, out of our worst dispositions," connects the blossoms of community, culture, and values back to their earthy roots in the individual psyche. In this vein, researchers have noted that the basic traits of the anal character are quite similar to many of the core values of Western civilization (Freud, 1930, 96–7). Because almost no anal eroticism is permitted in American culture, its repression is a powerful influence on personality and, therefore, community. Menninger (1943) considers the anal phase as "without doubt the most important period in the development of not only individuals but whole cultures as well." Researchers have expanded it to include Japanese culture and character (Spitzer, 1947; Gorer, 1943). Others have applied it to Hindu culture (Berkley-Hill, 1921). The anthropologist Roheim (1934, 283) suggests:

> From the frequency of anal characters in clinical analysis it was only one step to the conclusion that our culture or the type of our social organization in general was based on sublimations or reaction formations of anal trends. Where else can we find such an organized system of caring for the future, such as far-going possibility of "retaining" foodstuffs in a symbolic form? The infant receives nourishment from the mother and messes her with its excrements as a symbol of love. Civilization has invented a form in which the equivalent of nourishment can be retained for any length of time—money—and by means of which we can organize this retention—savings bank. No other group of human beings has such exaggerated ideas of hygiene and cleanliness, and whoever first said that the amount of soap used by a people was a sure indication of their level of "civilization" was not far from the truth. Soap, money, and regulations.

It is Freud (1930, 93) to whom Roheim refers:

> We do not think highly of the cultural level of an English country town in Shakespeare's time when we read that there was a big dungheap in front of his father's house in Stratford; we are indignant and call it "barbarous" (which is the opposite of civilized) when we find the paths in the Wiener Wald [Vienna Woods] littered with paper. Dirtiness of any kind seems to us incompatible with civilization. We extend our demand for cleanliness to the human body too. We are astonished to learn of the objectionable smell which emanated from the *Roi Soleil* [Louis XIV of France]; and we shake our heads on the [Isle of] Isola Bella when we are shown the tiny wash-basin in which Napoleon made his mourning toilet. Indeed, we are not surprised by the ideas of setting up the use of soap as an actual yardstick of civilization. The same is true of order . . . Order is a kind of compulsion to repeat which, when a regulation has been laid down once and for all, decides when, where, and how a thing shall be done, so that in every similar circumstance one is spared hesitation and indecision.

Research in sociology and anthropology lends support. Weber's (1930) seminal hypothesis that historically the rise of capitalism in the West rested upon the values and beliefs of Protestants, especially Calvinists, comports with the anal traits previously described. Those values included the virtue of hard work as such, the saving of money, a fetish for cleanliness and order, and a repressive, ascetic lifestyle. This connection, identified by many sociologists (Gerth and Mills, 1953; Erikson, 1958; Brown, 1959; Fromm, 1941), is summarized by Bocock (1976, 136):

> The Calvinist's distrust of the body and its sexuality ensured that there would be energy available to such a character for other activities in which this energy could find expression. In the specific case of anal erotism, the Calvinist's concern for care for detail in money matters, for saving and leading a frugal existence, and for bookkeeping can be seen as sublimated forms of this element of sexuality. The dedication to hard work was important in Weber's analysis of the growth of capitalism and science, entailing active mastery over the world by men . . .

So evident was its denial of anal and other sexual eroticisms, one early ascetic Protestant sect in North America was known as "Puritans," a connotation yet preserved in the meaning of "puritanical." Under capitalism in Colonial America, such anal traits and values found freedom and license. Over the centuries they crystallized into their present-day capitalist social structures, the superordinate institutions of business (making money), banking (hoarding money), and law (protecting money). To these the institutions of education, family, and religion remain subordinate and subservient (Scimecca, 1981).

In addition to the historical evidence, anthropology provides cross-cultural control groups which test the thesis. A contrast of our child-rearing practices to those in cultures in which repression at the anal stage of development is neglected or ignored, reveals, albeit imperfectly, its cultural and characterological importance for us. The Samoan culture, made famous by Margaret Mead (1928) and the central Australian cultures studied by Geza Roheim (1950) place no tabus or restrictions on excremental functions in child-rearing practices. Socially the anal phase of development is skipped or omitted in these societies. As we would expect, as a consequence, no anal characteristics develop. Time means nothing to these natives, there is a wholly lackadaisical attitude toward work, money is of little or no interest, and members share everything with their fellow beings. An emphasis on work and production, the value of time or the illusion that one can somehow "save" it, the importance of material possessions and hoarding, the striving for wealth and its implied power, all characteristic of civilization, are wholly foreign to them (Menninger, 1943). The social consequences, when contrast to our own customs and values, are stark:

> The . . . aborigines of Australia were able to store up enough food in advance, so that they could devote themselves to drama and dance for days, or even weeks on end . . . The warm, substantial bonds of kinship united man with man. There were no lords or vassals, serfs or slaves . . . No one kept another in bondage and lived upon the fruits of his labor . . . No one was ever sent to prison for stealing food to feed his children . . . There were no time clocks, no bosses or overseers . . . and a two week vacation was not one's quota of freedom for a year (White, 1959, 276–7).

A Caricature of Rural Community Values

The following abstraction, drawn from traits applicable to the individual (presented above), affords a clearer understanding of the connection of "our worst dispositions" to that which, as a culture, we trumpet as "our highest virtues."

The citizens are clean, orderly, obstinate, and independent-minded; they believe in individual rights and dignity, yet are dutiful to family and community; they do not like outsiders telling them what to do and do not like outside advice or interference in their affairs; they feel their self-control has been too much eroded by outside centralizing forces and too many "unclean" things are being imported into the community against their will—whether via magazines, cable television, or books in the public schools; they are irritated by such outside influences which they see as threatening "common human decency" and self-determination. Inflation, interest rates, unemployment, and crime mystify them and produce a generalized anxiety about the future. In the lower strata one will find a sense of helplessness accompanied by free-floating hostilities often expressed as religious or racial intolerance. They may overreact if they sense an injustice has been or will be done them, particularly if it pertains to their children whom they tend to over-glorify. They will take revenge and couch it in high moral terms. No one, in their opinion, knows what is better for them than they do. For some, if necessary, vigilanteism (sadism) is a distinct possibility. They like money and getting it is a positive dream to them, but strong residual values still center around family (children) and church. They hate waste and think of themselves as both clean and thrifty whether they are or not. In terms of community power centers, first-order power, the decision-makers, clearly resides in those institutions most intimately associated with anal traits and objects—business (making money), banking (hoarding money), and lawyers (orderliness and rule-bound mentality). Church-, school-, and civic-leaders are second-order decision-makers. In brief, we find they are "orderly, parsimonious, and obstinate."

In caricature, these are the basic contributions of anal stage repression to their core values. Other values important to siting, but not derived from the anal stage of development, the etiology of which is therefore outside the scope of this discussion, include a love of the environment (earth as female), a love of their country (father's land, authority), a belief in courts (a father-figure resolver of disputes), and a love of guns by males (symbol of masculine aggression and independence).

In the siting process, the two antithetical views of money—a good vs. an evil—may split a community. If a community's association of money with a sense of stain or bribery overrides its love of money, the usual first-order decision-makers—businessmen, bankers, and lawyers—all of whom have strong interest in money, may find themselves suddenly and uncharacteristically demoted or displaced in favor of church- and civic-leaders. Thus, despite the fact that site developers may have marshaled the support of the usual key leadership, resistance to waste siting may spring from unexpected sources such as the farm community, the church, civic organizations, or local representatives. Local politicians who attempt to straddle these shifting centers, of course, appear fickle.

Few of the attitudes and characteristics sketched above may be apparent in a rural community upon first encounter. They may be latent rather than manifest. These values are the core values which the community will fall back on (regress to) when stressed or threatened. It is on these they will take their stand. It is both their strength and their weakness, their vice and their virtue. Horowitz (1977, 45–6) summarizes succinctly:

> In the repression of anal erotism we can see all the sources of repression working concurrently: the ontogenetic recapitulation of the adoption of the erect posture by our pre-human ancestors, the maturational repression due to the progress of libido from zone to zone, the universal-cultural repression resulting from the requirement for cleanliness which must be present in any society, and the particular-cultural (surplus) repression which arises out of the puritan-capitalist

needs for supercleanliness, asceticism, orderliness, frugality, possessiveness, and competitiveness.

A SUPEREGO DYSTONIC: MORAL DEADLOCK

In connecting a hierarchy of dimensions via psychoanalytic theory—the biological (erotogenic stage of anal eroticism) to the psychological (anal personality as obstinate, thrifty, orderly), the psychological to the sociological (superordinate institutions such as banking and law), and the sociological to the anthropological (cultural values such as individualism, a belief in rights, love of money, etc.)—we have found that a major part of the root cause of community resistance is earthy indeed—*ourselves.* Far from being irrational, this resistance springs from who we are, how we are made, how our id was transformed into ego and superego and stamped "Made in America."

Americans participate in a predominately anally-fixated society as evidenced by many of its most distinguishing cultural traits and values, values which are either anal retentive displacements (a fetish for hoarding money, making money, thrift, concern about time, "time is money," etc.) or anal retentive reaction formations (a fetish for cleanliness, order, regulation, domination, and control). Its dominate powerholders (businessmen, bankers, and lawyers) are the functional representatives of basic anal character traits which, across centuries, have been institutionalized and crystallized into social structure and values.

Our cultural identity, our core values, and our superordinate institutions represent a formation of reactions against "dirt" in any symbolic form. With respect to waste siting, the origin of our personal resistance to our own wastes is the origin of a community's collective resistance to its return. *We cannot accept the return of that which we had to repudiate in order to become civilized.* Just as ego cannot easily undo its repressions (Freud, 1926, 153), neither can a community. To do so would be for almost any community, as it would for a normal individual, psychologically self-destructive.

These links are an insight alarming and profound. For if this analysis has merit, then the problem of waste siting is more serious than previously supposed. Hazardous or radioactive waste disposal is not just another social problem in society, but one marked by a resistance which cuts across traditional social categories and theoretical perspectives and which operates consciously or unconsciously at all levels. It—the return of the repressed—is a conflict which rends at the core of identity.

Usually self-delusion ("lying honestly") has some positive benefit; lying to ourselves helps us get a job done. But not in this instance, not on this issue. Objective reality will not permit it and the unconscious will not accept it. There is no point in attempting to deny or sanitize the waste disposal problem by telling ourselves a lie which we cannot possibly believe. Whether we consciously want to admit it or not, hazardous waste represents the return of repressed material from the unconscious (id). It is the absolutely forbidden, the antithesis of what it means to be civilized. Waste-filth-dirt-stain symbolically, as well as organically, threatens to break down all that which, by its very repression, we have labored to build up and cherish—the core of our personalities, our community, our culture, our civilization. This repression constitutes an essential part of our sense of physical, communal, and spiritual cleanness.

In a sociological sense, society appears to be moving toward a collision against the core of its own deepest values. What Freud (1905, footnote 1920, 187n) wrote with respect to a child's "first repression," quoted earlier, we now apply to that which, in order to bring itself into being and sustain itself, becomes civilization's foremost repression as well:

> From that time on, what is 'anal' remains the symbol of everything that is to be repudiated and excluded from life.

A Policy Recommendation: Remote Removal

In its 1983 review of the national hazardous waste problem, the U.S. Office of Technology Assessment (OTA) painted an ominous picture. Disposal of such wastes over a number of decades has been largely unregulated; some 80,263 sites exist with contaminated surface impoundments, 90 percent of which pose a potential threat of ground-water contamination; the long-term health threats of hazardous waste are uncertain and potentially serious; and economic costs of managing hazardous wastes are estimated at $4–5 billion and rising rapidly. In 1985 OTA found that, despite expenditure of $1 billion from the Superfund, only 30 percent of the 538 sites on the U.S. Environmental Protection Agency's priority list were receiving remedial cleanup attention and that such remedial cleanups tend to be impermanent, some sites get worse and repeated costs are inevitable. In practice, risks are merely transferred from one community to another and to future generations (Kasperson,

1985, 10). In March, 1985, the OTA estimated that more than 10,000 disposal sites for hazardous waste will require clean-up at a cost which may eventually total $100 billion. At the same time, the ongoing generation of hazardous wastes will require successfully siting hundreds of new facilities across the country (Morell, 1984). To this picture may be added the ubiquitous problems of solid wastes and the special problems of low-level and high-level radioactive wastes.

As a culture, present policy represents a form of slow self-poisoning. We continue to limp down a tortuous path destined to befoul environmentally and psychologically a once pristine continent, turning it into a chemical ghetto. The latest fixes of techno-rationality—new liners, new foams, new concrete bunkers, sophisticated remote monitoring systems—are not the answer for a superego dystonic. Waste siting city-by-city, county-by-county, state-by-state, region-by-region is not a solution. It is the problem—a mistake which we apparently have a compulsion to repeat. Our institutional behavior and goals with respect to waste management are fundamentally incompatible with our cultural ideals, self-conception, and emotional well-being.

The problem is intractable, but one option does exist. Avoid it. Keep all forms of waste away from people, out of our communities, off our farm land, away from our aquifers, and out of our watersheds. Conceive each of the separate waste streams which now exist as components of an integrable whole for the purpose of creating one national all-encompassing waste management system. Take all nonorganic wastes and, to the extent appropriate, incinerate them, dry them, sort them, compact them, barrel them, compact them again, containerize them, consolidate them, unit-train them, and bury them in an arid, uninhabited area on federal land in the desert, such as those vast regions already "crapped up" and sacrificed by nuclear weapons testing.

CONCLUSION

Wrote Freud (1923, 48–9), "The superego is always close to the id and can act as its representative *vis-a-vis* the ego. It reaches deep down into the id and for that reason is farther from consciousness than the ego is." In this paper the psychoanalytic perspective reaches down into the psyche to appreciate the sense in which hazardous wastes and their inevitable mismanagement do not offend the environment or nature nearly so much as they offend our culture and our souls. Our dead ancestors, in building this civilization, left us a colloquial commandment derived from the bitter lessons taught them by decimating diseases and other disciplines of the reality principle: Cleanliness is next to godliness. This is not mystical or religious, but deep psychology. It is the cryptic language in which they summarized for us the interconnection of the dangers, bottom to top, of the Waste Itself Issue.

Those responsible for managing and disposing of society's wastes should set a higher course, one which sails against the institutional eddies of the present and with the cultural wind for a better future. The problem is real, the task Herculean, but the work is worthy of high self-esteem. In psychoanalytic terms, the plaque upon the wall of those broad-visioned leaders, should they emerge, who essay the nation's problem of waste disposal beyond the din of provincialism and into a unified system dedicated to remote removal, should read:

> When the ego has brought the super-ego the sacrifice of an instinctual renunciation, it expects to be rewarded by receiving more love from it. The consciousness of deserving this love is felt by it as pride (Freud, 1939, 117).

REFERENCES

The date given for Freud's works is the year the work was originally published. Freud's works may be found in the 24-volume *The Standard Edition of the Complete Psychological Works of Sigmund Freud* as cited. Page number references in the text are to *The Standard Edition.*

Abraham, Karl: A short study of the development of the libido. In *Selected Papers.* London, Hogarth Press, 1924.

Adorno, T. W., Frenkel-Brunswik, E., Levinson, D. J. and Sandford, R. N.: *The Authoritarian Personality.* New York, Harper & Row, 1950.

Altheide, David L. and Johnson, John M.: *Bureaucratic Propaganda.* Boston, Allyn and Bacon, 1980.

Anonymous: Radiation waste site not wanted. *Daily News.* Effingham, Illinois, August 18, 1987.

Berkley-Hill, O.: The anal erotic factor in religion, philosophy and character of the Hindus. *International Journal of Psychoanalysis, 2:*306, 1921.

Blackburn, Thomas: Personal communication. Texas Low-Level Radioactive Waste Authority, 1985.

Blatt, S. J.: An attempt to define mental health. *Journal of Consulting Psychology, 28:*146–153, 1964.

Bloom, Allen: *The Closing of the American Mind.* New York, Simon and Schuster, 1987.

Bocock, Robert: *Freud and Modern Society: An Outline and Analysis of Freud's Sociology.* London, Thomas Nelson and Sons, 1976.

Bourke, John G.: *Scatalogic Rites of All Nations.* Washington, D. C., W. H. Lowdermilk, 1891.

Brown, N. O.: Studies in anality. *Life Against Death.* London, Rutledge & K. Paul, 1959.

Campos, L. P.: Relationship between time estimation and retentive personality traits. *Perceptual and Motor Skills, 23:*59–62, 1966.

Colglazier, William E.: Ethical and value issues in radioactive Waste Management. MSS. Knoxville, The University of Tennessee, 1986.

Durkheim, Emile: *The Rules of Sociological Method.* New York, Free Press, 1895 (English, 1964).

Emerson, Ralph Waldo: *The Conduct of Life: Nine Essays (Worship),* Boston, Ticknor and Fields, 1860.

Erikson, E.: *Young Man Luther.* New York, Norton, 1958.

Fenichel, O.: *The Psychoanalytic Theory of Neurosis.* New York, Norton, 1945.

Finney, J. C.: The MMPI as a measure of character structure as revealed by factor analysis. *Journal of Consulting Psychology, 25:*327–336, 1961.

Fisher, Seymour and Greenberg, Roger P.: (1977) *The Scientific Credibility of Freud's Theories and Therapy.* New York, Basic Books, 1977.

Fox, Robin: Totem and taboo reconsidered. In Leach, E. (Ed.): *The Structural Study of Myth and Totemism.* London, Tavistock Publications, 1967.

Freud, Anna: *The Ego and the Mechanisms of Defense.* London, Hogarth Press and the Institute of Psychoanalysis, 1937 (English, 1948).

Freud, Sigmund: Further remarks on the neuro-psychoses of defence. 1896. In Strachey, James (Ed.): *The Standard Edition of the Complete Psychological Works of Sigmund Freud,* London, Hogarth Press, 1955–1968, vol. III.

Freud, Sigmund: *The Interpretation of Dreams.* 1900. In Strachey, James (Ed.): *The Standard Edition of the Complete Psychological Works of Sigmund Freud,* London, Hogarth Press, 1955–1968, vol. V.

Freud, Sigmund: *Three Essays on the Theory of Sexuality.* 1905. In Strachey, James (Ed.): *The Standard Edition of the Complete Psychological Works of Sigmund Freud,* London, Hogarth Press, 1955–1968, vol. VII.

Freud, Sigmund: Character and anal erotism. 1908. In Strachey, James (Ed.): *The Standard Edition of the Complete Psychological Works of Sigmund Freud,* London, Hogarth Press, 1955–1968, vol. IX.

Freud, Sigmund: Formulations on the two principles of mental functioning. 1911. In Strachey, James (Ed.): *The Standard Edition of the Complete Psychological Works of Sigmund Freud,* London, Hogarth Press, 1955–1968, vol. XII.

Freud, Sigmund: Preface to Bourke's *Scatalogic Rites of All Nations.* 1913. In Strachey, James (Ed.): *The Standard Edition of the Complete Psychological Works of Sigmund Freud,* London, Hogarth Press, 1955–1968, vol. XII.

Freud, Sigmund: The claims of psychoanalysis to scientific interest. 1913a. In Strachey,

James (Ed.): *The Standard Edition of the Complete Psychological Works of Sigmund Freud,* London, Hogarth Press, 1955–1968, vol. XIII.

Freud, Sigmund: Repression. 1915. In Strachey, James (Ed.): *The Standard Edition of the Complete Psychological Works of Sigmund Freud,* London, Hogarth Press, 1955–1968, vol. XIV.

Freud, Sigmund: *Introductory Lectures on Psychoanalysis.* 1916–1917. In Strachey, James (Ed.): *The Standard Edition of the Complete Psychological Works of Sigmund Freud,* London, Hogarth Press, 1955–1968, vol. XV.

Freud, Sigmund: *Group Psychology and the Analysis of the Ego.* 1921. In Strachey, James (Ed.): *The Standard Edition of the Complete Psychological Works of Sigmund Freud,* London, Hogarth Press, 1955–1968, vol. XVIII.

Freud, Sigmund: *The Ego and the Id.* 1923. In Strachey, James (Ed.): *The Standard Edition of the Complete Psychological Works of Sigmund Freud,* London, Hogarth Press, 1955–1968, vol. XIX.

Freud, Sigmund: The resistances to psycho-analysis. 1925. In Strachey, James (Ed.): *The Standard Edition of the Complete Psychological Works of Sigmund Freud,* London, Hogarth Press, 1955–1968, vol. XIX.

Freud, Sigmund: *Inhibition, Symptoms and Anxiety.* 1926. In Strachey, James (Ed.): *The Standard Edition of the Complete Psychological Works of Sigmund Freud,* London, Hogarth Press, 1955–1968, vol. XX.

Freud, Sigmund: *Civilization and Its Discontents.* 1930. In Strachey, James (Ed.): *The Standard Edition of the Complete Psychological Works of Sigmund Freud,* London, Hogarth Press, 1955–1968, vol. XXI.

Freud, Sigmund: New introductory lectures on psycho-analysis. 1933. In Strachey, James (Ed.): *The Standard Edition of the Complete Psychological Works of Sigmund Freud,* London, Hogarth Press, 1955–1968, vol. XXII.

Freud, Sigmund: Construction in analysis. 1937. In Strachey, James (Ed.): *The Standard Edition of the Complete Psychological Works of Sigmund Freud,* London, Hogarth Press, 1955–1968, vol. XXIII.

Freud, Sigmund: An outline of psychoanalysis. 1938. In Strachey, James (Ed.): *The Standard Edition of the Complete Psychological Works of Sigmund Freud,* London, Hogarth Press, 1955–1968, vol. XXIII.

Freud, Sigmund: *Moses and Monotheism.* 1939. In Strachey, James (Ed.): *The Standard Edition of the Complete Psychological Works of Sigmund Freud,* London, Hogarth Press, 1955–1968, vol. XXIII.

Fromm, E.: *Fear of Freedom.* New York, 1941.

Gerth, Hans and Mills, C. Wright: *Character and Social Structure.* New York, Harcourt, Brace, 1953.

Gorer, G.: Themes in Japanese culture. *Transactions of the New York Academy of Science, 2:*106–124, 1943.

Gross, Bertram: *Friendly Fascism.* New York, M. Evans, 1980.

Hensler, Deborah R. and Hensler, Carl P.: *Evaluating Nuclear Power: Voter Choice on the California Nuclear Energy Initiative – Executive Summary.* July, 1979.

Horowitz, Gad: *Repression: Basic and Surplus Repression in Psychoanalytic Theory.* Buffalo, University of Toronto Press, 1977.

Jacobs, M. A., Knapp, P. H., Anderson, L. S., Karush, N., Meissner, R., and Richman, S. J.: Relationship of oral frustration factors with heavy cigarette smoking in males. *Journal of Nervous and Mental Disease, 141:*161–171, 1965.

Jacobs, M. A., Anderson, L. S., Champagne, E., Karush, N., Richman, S. J., and Knapp, P. H.: Orality, impulsivity and cigarette smoking in men: further findings in support of a theory. *Journal of Nervous and Mental Disease, 143:*207–219, 1966.

Jacobs, M. A., and Spilken, A. Z.: Personality patterns associated with heavy cigarette smoking in male college students. *Journal of Consulting and Clinical Psychology, 37:*428–432, 1971.

Jacoby, Russell: *Social Amnesia.* Boston, Beacon Press, 1975.

Jones, E.: Anal erotic character traits. In *Papers on Psychoanalysis.* London, Bailliere, Tindall & Cox, 1923.

Jung, C. G.: *Studies in Word Association.* London, Heinemann, 1918.

Kasperson, Roger E. (Ed.): *Equity Issue in Radioactive Waste Management.* Cambridge, Massachusetts, Oelgeschlager, Gunn & Hain, 1983.

Kasperson, Roger E.: Rethinking the siting of hazardous waste facilities. Unpublished manuscript, Worcester, Massachusetts, Center of Technology, Environment, and Development, Clark University, 1985.

Kline, Paul: *Fact and Fantasy in Freudian Theory.* London, Methuen, 1972.

Lake, Robert W.: *Resolving Locational Conflict.* New Brunswick, New Jersey, Center for Urban Policy Research, 1987.

Lasch, Christopher: *The Culture of Narcissism.* New York, W. W. Norton, 1978.

Lerner, B.: Auditory and visual thresholds for the perception of words of anal connotation: an evaluation of the "sublimation hypothesis" on philatelists. Doctoral dissertation, Yeshiva University, 1961.

Mackinnon, D. W., and Dukes, W. F.: Repression. In Postman, L. (Ed.): *Psychology in the Making.* New York, Knopf, 1962.

Marcuse, Herbert: *One Dimensional Man.* Boston, Beacon Press, 1964.

Mead, Margaret: *Coming of Age in Samoa.* New York, Wm. Morrow, 1928.

Menninger, William C.: Characterologic and symptomatic expressions related to the anal phase of psychosexual development. *Psychoanalytic Quarterly, 12:*161–193, 1943.

Morell, David, and Magorian, Christopher: *Siting Hazardous Waste Facilities: Local Opposition and the Myth of Preemption.* Cambridge, Massachusetts, Ballinger, 1982.

Morell, David: Siting and the politics of equity. *Hazardous Waste, 1:*(No. 4), 555–571, 1984.

Noblin, D. D.: Experimental analysis of psychoanalytic character types through the operant conditioning of verbal responses. Doctoral dissertation, Louisiana State University, 1962.

O'Hare, Michael, Bacow, Lawrence, and Sanderson, Debra: *Facility Siting and Public Opposition.* Cincinnati, Van Nostrand, Reinhold, 1983.

Otway, Harry, and Thomas, Kerry: Reflections on risk perception and Policy. *Risk Analysis, 2:*69–82, 1982.

Pettit, T. F.: Anality and time. *Journal of Consulting and Clinical Psychology, 33:*170–174, 1969.

Roheim, Geza: The study of character development and the ontogenetic theory of culture. *Essays Presented to C. G. Seligman.* London, Kegan Paul, Trench & Trubner, 1934.

Roheim, Geza: *Psychoanalysis and Anthropology.* New York, International Universities Press, 1950.

Rosenwald, G. C.: Effectiveness of defenses against anal impulse Arousal. *Journal of Consulting and Clinical Psychology,* 39, 292–298, 1972.

Rycroft, Charles: *A Critical Dictionary of Psychoanalysis.* Harmondsworth, Penguin, 1972.

Sandler, J. and Hazari, A.: The obsessional: on the psychological classification of obsessional character traits and symptoms. *British Journal of Medical Psychology, 33:*113–121, 1960.

Sandman, Peter M.: Getting to maybe: some communications aspects of siting hazardous waste facilities. In Lake, Robert W. (Ed.): *Resolving Locational Conflict.* New Brunswick, New Jersey, Center for Urban Policy Research, 324–344, 1987.

Schlesinger, V. J.: Anal personality traits and occupational Choice: A study of accountants, chemical engineers and educational psychologists. Doctoral dissertation, University of Michigan, 1963.

Schwartz, Steven P., White, Paul E., and Hughes, Robert G.: Environmental threats, communities, and hysteria. *Journal of Public Health Policy, 6:*58–77, 1985.

Scimecca, Joseph A.: *Society and Freedom: An Introduction to Humanist Sociology.* New York, St. Martin's Press, 1981.

Segal, S. J.: A psychoanalytic analysis of personality factors in vocational choice. *Journal of Counseling Psychology, 8:*202–210, 1961.

Skinner, B. F.: *Beyond Freedom and Dignity.* New York, Knopf, 1971.

Sorokin, Pitirim A.: *Society, Culture, and Personality.* New York, Harper, 1947.

Spitzer, H. M.: Psychoanalytic approaches to the Japanese character. *Psychoanalysis and the Social Services,* New York, International Universities Press, 1947, vol. 1.

U.S. Environmental Protection Agency: *Draft Environmental Impact Statement for Subtitle C.* Washington, 1979.

U.S Environmental Protection Agency: *Siting of Hazardous Waste Management Facilities and Public Opposition.* Washington, D.C., Centaur, 1979a.

U.S. General Accounting Office: *Siting of Hazardous Waste Landfills and their Correlation with Racial and Economic Status of Surrounding Communities.* Washington, D.C., 1983.

U.S. Office of Technology Assessment: *Technologies and Management Strategies for Hazardous Waste Control.* Washington, USGPO, 1983.

von Winterfeldt, Detlof, and Edwards, Ward: Patterns of conflict about risky technologies. *Risk Analysis, 4:*55–68, March, 1984.

Weber, Max: *The Protestant Ethic and the Spirit of Capitalism.* New York, Scribner, 1930.

Weber, Max: *Theory of Social and Economic Organization.* New York, Oxford University Press, 1947.

White, Leslie: *The Evolution of Culture.* New York, McGraw-Hill, 1959.

Chapter 13

THE TECHNOLOGICAL WORLDVIEW AND ENVIRONMENTAL PLANNING

JOHN W. MURPHY

At this time technology is proliferating at an unprecedented rate. Related to this growth are negative social implications, one of which pertains to the disposal of toxic waste. Although many recent technological developments appear to be destroying the environment, reversing this trend seems to be nearly impossible. In fact, the problems that result from technology are usually corrected through the creation of additional technical knowledge. Consequently, modern society is trapped within a spiral of ever increasing technology. Before environmentalists can be taken seriously, however, this cycle must be broken. The tunnel vision imposed by technology must be overcome.

Overlooked is the method whereby technology perpetuates itself (Leiss, 1976, pp. 113–124). Specifically, technology is most often treated as a tool that has a wide range of applications, and therefore its use is assumed to be naturally expanding. A variety of theorists and technicians, on the other hand, argue that technology is anything but a tool (Murphy, Mickunas, and Pilotta, 1986). Technology is understood to be accompanied by a worldview that shapes how persons perceive themselves and their surroundings. Actually, these underlying principles make the use of technology possible. Once this worldview is adopted, the likelihood is increased that technical knowledge will be assessed as valuable. Reality, simply put, is defined in a manner that encourages technical solutions to social problems. Yet the impact the technological worldview may have on the formulation of environmental policies has generally been ignored. The thrust of technology, in short, is not technological but philosophical.

To better appreciate the following discussion, the reader must entertain the idea that reality is not obvious. Following the scientific work of Heisenberg, Planck, and Gödel, not to mention literary critics and philosophers such as de Man, Derrida, and Foucault, ushered in has been the Postmodern Era (Lyotard, 1984). Central to postmodernist position is that reality is a product of "language games," and, thus, knowledge cannot be associated simply with obtrusive empirical referents. Instead, reality is based on the acceptance of certain assumptions about the nature of knowledge and order. These themes, accordingly, are deployed through linguistic acts. As Berger and Luckmann (1966) contend, reality is "socially constructed." In the Postmodern Era, according to Jacques Lacan (1977, p. 306), truth is derived from language and not reality. Essential to the success of technology, therefore, is that symbols are manipulated in a particular way.

William Leiss (1974, pp. 145–165) suggests that the modern world is dominated by the technological worldview. The mode of cognition conveyed by technology has been accepted unquestioningly as representing reality. A sort of technological hegemony exists. What this means is that the utility of technology is predicated on a type of self-fulfilling prophecy. In order to implement technology successfully, particular reality assumptions must be accepted while others are rejected. These underlying beliefs, moreover, encourage the use of technology. The resulting social imagery has been referred to by Marcuse (1964) as "one dimensional." He is suggesting that because a specific understanding of reality is presumed to be ultimately rational, all other viewpoints are automatically discredited and dismissed as unsuitable to sustain public policy. As might be suspected, subsequent to this démarche persons are easily convinced that only one way of approaching a problem is valid. This is how the utilization of technology is both facilitated and perpetuated.

Technology, therefore, cannot necessarily be controlled by technocrats. In fact, technical knowledge merely insures the survival of technology. This idea is not novel, but one that has not been sufficiently exploited. While describing what he believed to be the key crisis facing European society, Edmund Husserl (1970) recognized that tempering the impact of technology is not a technical issue. Much more important, a variety of philosophical and theoretical questions must be addressed before technology will enhance, rather than depreciate, the human condition. In other words, unless the conceptual side of technology is thoroughly examined, the myriad of devices that is supposed to improve the lives of persons may become their adversary.

THE TECHNOLOGICAL WORLDVIEW

In a classic statement, Lewis Mumford (1963) wrote that control is the most valuable derivative of technology. Technology is able to control events more effectively than ever before imagined. Jacques Ellul (1964, p. 134) claims that such control is possible because a "technological ethic" guides the implementation of technology. Simply put, technocrats believe that every activity can be legitimately transformed into technical operations, without any loss of meaning. With the human or interpretive element removed as much as possible from any task, the error usually associated with volition can be minimized. Some modern commentators charge that today society functions on the basis of a "technological imperative," because whenever possible technology is supposed to be introduced to insure that a job is sufficiently standardized (Schuurman, 1980). As a result of the formalization that is attendant to the use of technology, Don Ihde (1982) notes that practically any setting can be successfully "deanimated."

The "emotional control" that Peter Berger and his colleagues (1973) state is provided by technology is the product of a unique "language game." When the world is mediated by technology, reality is misconstrued to be dualistic. Specifically, a significant distance is interjected between a knower and whatever is known. By doing this, the belief is reinforced that knowledge uncontaminated by interpretation can be acquired. As described by the early Greeks, *doxa*, or opinion, can be kept from impeding the discovery of facts. Most interesting about the technological world-view, the illusion is promoted that language is unrelated to reality. In other words, persons are systematically alienated from their milieu through technology. Increased precision is likely, yet a high price is paid for this improvement.

The purpose of technology is to neutralize everything, and the environment is no exception. Perpetuated by the use of technology is dualism, and thus interpretation is thought to be disassociated from any technical undertaking. Reason guides policy implementation, rather than capriciousness. Through the use of technology the world is ostensibly rendered available for study and expropriation. In a high-tech world questions of morality are transformed into technical operations. Hence technical feasibility statements determine the worth of a project. In this regard, technology begins to dominate its users.

The technological worldview consists of three different, but interrelated, principles (Murphy and Pardeck, 1986). First, all phenomena are materialized, or envisioned to be inert objects. Second, quantitative indices are thought to be most appropriate for supplying events with an identity. And finally, the laws of physics are assumed to cause both social and natural occurrences. Using the phrase popularized by Martin Heidegger (1977, p. 17), the world becomes a "standing reserve," due to the theoretical tenets that are pivotal to the use of technology. Society is portrayed as a domain filled with numerous resources, which can be extracted in any manner that is deemed technologically feasible. Consequently, the land is allowed to be completely derogated!

Every phenomenon is assumed to be empirical, and thus free of values. Further, the entire social scene is purged of symbolism, because persons are conceived to be simply material objects. The environment is not intimately connected to human values, but is a source of riches that does not have to be respected. For the notion of respect presupposes the legitimacy of judgments. Following the subversion of humanly inspired ideals, only technological principles remain to direct the use of technology. A technological animus thus begins to dictate all development.

Readers should not be surprised when Ellul (1984) declares that the hallmark of the modern world is "technological slavery." Because persons have abandoned their values in order to be scientific, technology cannot be constrained by social commitments. Whatever is technologically possible, accordingly, is normally undertaken. All that will curtail a project are technical difficulties. Because political, ethical, and aesthetic concerns are anathema to technological rationality, issues such as these are considered to be disruptive. Social planning, therefore, is nothing more than instrumental activity, whereby the worth of a particular course of action is measured solely according to technical criteria. Obviously, sentimental attachment to the land is of little concern to a technocrat.

THE COMPUTER MICRO-WORLD: AN EXAMPLE OF THE TECHNOLOGICAL WORLDVIEW

Many computer theoreticians do not believe that computers are simply machines. Joseph Weizenbaum (1976, p. 74), the creator of the program ELIZA, maintains that computers should not be imagined to be grand calculators. And, Margaret Boden (1977, pp. 15–17) argues that computers have the ability to manipulate symbols, and thus do not simply "crunch" numbers. Presupposed by the use of computers are specific epistemological parameters that must be endorsed, or this technology will have little utility. Without adopting a particular "frame" for representing knowledge, Marvin Minsky (1981) demonstrates that information is unable to be processed by a computer. Minsky and Seymour Papert (1971) pursue this point further and write that computers operate on the basis of a unique "micro-world."

In a manner of speaking, input is perceived and evaluated by computers. A computer program is comprised of categories, arranged like the mailboxes found in the vestibules of apartment buildings, into which data are inserted. Indeed, these slots are often referred to as "address spaces," and, in fact, should be treated as analogous to cognitive categories or conceptual digits (Bolter, 1984, pp. 83–90). Before data can be entered into a computer, however, they must be able to be neatly classified. Knowledge that is amorphous and readily interpreted in any number of ways will be rejected by a computer. For this reason, writers such as Jean-Francois Lyotard (1984, p. 4) contend that synonymous with the so-called Computer Age is the information "bit."

Conceived to be a bit, knowledge can be isolated and classified. This reason for this is that input is understood to have precise dimensions. Ambiguity is eliminated as a result of adopting an esoteric form of reasoning, sometimes referred to as binary logic. Simply put, A is thought to be equal to A, B = B, and A ≠ B. Suggested by this symbolism is that events have a single identity, and thus are stripped of their interpretive significance (Dreyfus, 1979, pp. 70–75). When transformed into something objective, a phenomenon can be categorized without much difficulty. Yet destroying the human character of knowledge can be extremely problematic, especially when social data are the focus of attention.

Those who use computers are socialized into adopting binary or formal logic, the cornerstone of the computer micro-world. In addition to mastering certain technical intricacies, students and other computer operators are sent a clear message. That is, binary logic represents the paragon of reason. Due to the dualism that is inherent to this style of reasoning, symbols are construed to "represent elementary facts about the world" (Dreyfus and Dreyfus, 1986, p. 53). Additionally, interpretation is assumed to prevent a person from gaining access to reality. In order for truth to be obtained, passion must be expelled from the knowledge acquisition process. Those who use computers are told implicitly that logistical rigor and precision guarantee the generation of valid knowledge.

From the foregoing discussion, readers should begin to appreciate what Boden means when she says that symbols are manipulated by computers. Clearly, through computerization reality is defined in technical terms, thereby narrowing a person's field of vision. At first the development of contextless knowledge was considered to be a breakthrough, yet now this is not assumed to be the case. Knowledge divorced from the human condition has little relevance for predicting events that are affected by situational exigencies. And because persons cannot be separated from their environment, contextless knowledge is useless for conceptualizing social behavior. In short, the computer micro-world conveys an idealized image of knowledge. As with the technological world-view in general, the computer micro-world unduly "rationalizes" or "formalizes" social life.

IMPLICATIONS OF THE TECHNOLOGICAL WORLD-VIEW

What does it mean to say that the modern world is thoroughly rationalized? When existence becomes rationalized, as Max Weber (1978, pp. 85–86) points out, substantive rationality is gradually removed from the world. Specifically, values, beliefs, and commitments are excluded systematically from assessments and all other aspects of policy formation. The world is thus treated as a reservoir to be utilized in the most efficient manner possible. In the absence of other values, the value base of technology begins to determine the course of social development. What are some of the ways in which this occurs?

1. Knowledge is treated as if it is basically empirical. What this means is that so-called objective properties sustain all judgments. If objectively the environment can tolerate the creation of another nuclear power plant, then most likely such a project will be inaugurated. Most devastating about this rendition of knowledge is that nonscientific data are cited to impede progress. The fears that parents may have about the safety of their children, in case of a radiation leak, are not usually considered to be sufficient to prevent the opening of a nuclear station. Only "hard" data are important enough to be introduced regularly into the process whereby the future of the environment is debated. Emotional issues are not believed to be worthy of serious consideration. In fact, subjective insight is presumed to stifle rational discussion.

2. Sources of valid knowledge are typically associated with certain privileged segments of society. Usually technocrats, those educated professionals who extoll the benefits of science, are recruited to sanction policies. Recent research demonstrates, however, that these experts study environmental issues with a particularly narrow mind-set (Hisker, 1986). Indeed, only a limited number of options are given credence, while other possibilities are flippantly dismissed. True public debate is impossible, for input is not sought from a wide range of persons. Technocrats do not consider policy making to be a democratic undertaking.

3. Restricted are the ways in which data are collected and interpreted. A favorite method of policy makers is cost-benefit analysis (Murphy, 1985). Accordingly, data are given a monetary value and economic reasoning is used to generate all findings. Elaborate models are available that allow a plethora of variables to be included in any calculation. Nonetheless, the so-called "intangibles" associated with noneconomic factors are excluded from an assessment. Even the social cost-accounting procedures that are sometimes used are patently reductionistic. Before a variable can receive any serious attention it must be translatable into economic terms. Due to this methodology, many aspects of life that cannot be easily quantified are overlooked.

4. Accompanying this style of empiricism is unique social imagery, usually labeled atomism. In order to produce neat calculations, social life must be divided into manageable units. Once the various components of society have been isolated and given an explicit value, each one can be accurately evaluated. Obscured by these measurements, however, are the relationships among environmental factors. Most modern critics recognize that a holistic outlook is vital for conducting a useful study of society, particularly of the environment (Gorz, 1980, pp. 3–10). Additionally, the importance of change is minimized by empiricism, and thus the research that is inaugurated tends to emphasize the static nature of reality. Nature does not make leaps, charge empiricists. Yet the environment is dynamic and should not be treated as something that can be dissected, without destroying its integrity.

According to Habermas (1973, pp. 268–282) and other Frankfurt theorists, technology has become an ideology. What they mean by this derogatory appellation is that the knowledge indigenous to the technological worldview is claimed to be beyond reproach and thus constitutes a reality *sui generis.* Because technical knowledge is unencumbered by interpretation, *techné* is believed to have universal appeal. Anyone who uses technology is expected to be unbiased, and therefore can be trusted to make judgments that advance the common weal. Yet, as noted in the foregoing discussion, in this instance value freedom is a chimera, for subtending the use of technology is technological rationality. Most diabolical about an ideology is that values are concealed which are actually shaping reality. Hence the values that subtend technology are obscured.

As a result of the value base that accompanies technology, the range of possible social discourse is unnecessarily truncated. Only certain questions can be raised and answers posed. Specifically important is that any serious discussion of political, ethical, and moral principles is eschewed, because issues such as these may distort a person's view of empirical reality. Allegedly these are "subjective" concerns. In this way, the dream of capturing knowledge that is unsullied by human touch is preserved. The *doxa* that has perennially haunted the West is finally vanquished. However, this triumph is not without a price, in that the environment becomes an objective and a lifeless region.

SOCIALLY RESPONSIBLE TECHNOLOGY

The focal point of the technological worldview is dualism. Basic to this philosophical tenet is that specific knowledge is protected from the influence of interpretation. This pristine information is given a seignorial status, while so-called subjective insight is treated as suspect. Technologically-based knowledge is thus assumed to be factual, and more valuable than any competing mode of understanding.

In order for technology to be socially responsible, dualism must undermined. Deprived of a dualistic foundation, technology is not necessarily alienating. Modern writers refer to this nondualistic setting as a "reflexive environment" (Murphy, Mickunas, and Pilotta, 1986). As a result of creating this type of context, the rules for implementing technology are derived from the human *Lebenswelt,* or "life world." In other words, the principles that underpin technology are viewed to represent an interpretation of reality. Technological reason is merely one mode of conceptualization among others. Accordingly, technology that is not alienating is certainly conceivable. That is, tempering the influence of technology by other forms of knowledge will not necessarily undermine rational planning. The following points are provided to justify this conclusion.

1. As demonstrated by postmodernists such as Lyotard, Jacques Derrida,

and Paul de Man, for example, the idea of autonomous knowledge can no longer be sustained (Melville, 1986). The reason for this is that dualism cannot be theoretically justified. Dualism has been rejected because language is recognized to be at the center of the knowledge acquisition process. While borrowing from Wittgenstein, postmodernists charge that all knowledge originates from "language games," even the principles that legitimize the use of science and technology. Consequently, every style of knowledge is mediated fully by interpretation, while technical reason represents one way of conceptualizing social life. Subsequent to this philosophical démarche, technology is recognized to be a part of the "life-world" that is deployed through creative *praxis*. Technology is not a reality *sui generis*, but has utility only when certain assumptions about the world are accepted. Hence, technological rationality has a contingent status, similar to so-called mundane modes of assessment.

2. Because knowledge is linguistically instituted, community-based strategies must be employed when studying society (Murphy and Pilotta, 1983). Specifically, methods should be used that are sensitive to the interpretive character of reality. Certain knowledge, simply because it is touted to be scientific, should not be collected and included automatically in the formulation of policies. Likewise, so-called nonscientific information should not be dismissed as if it is unimportant by nature. Since all knowledge is replete with the "undecidables" associated with interpretation, so-called scientific facts are not inherently more valuable than any other data. Hence policies should be based on information that is socially meaningful, as opposed to knowledge that is recommended by a coterie of experts. Yet the opinions held by the citizenry are not always assessed by technocrats to be germane to developing sound policy proposals.

3. Order and social relations among persons must also be treated as fundamentally linguistic. Accordingly, the status distinctions that exist among persons are symbolic. In the case of technocrats and citizens, neither of these two groups is necessarily superior to the other. But if the value base of technology is given primacy in a society, then the former group is given immense power. In a pluralistic society, contrary to a bureaucracy, interpersonal discourse is supposed to be symmetrical. When this form of interaction is encouraged an argument is judged on its merits, and an idea is not accepted or rejected simply because it emanates from a particular segment of society. Jürgen Habermas (1970) maintains that before such "nonrepressive" discourse is possible, order

must be acknowledged to develop from diverse values, as opposed to an "ultimate reality." When based on the acceptance of diversity, public hearings can be organized to guarantee that as many citizens as possible participate in discussions. In this way, policies are an extension of community action and their formation is democratized.

4. The utilization of technology must be understood as value-based. Therefore, alleged progressive thinking is not necessarily exemplified through the exercise of technology. If realism, efficiency, and quantification are important, then the production and dissemination of technology will probably proceed unabated. On the other hand, in a situation dominated by aesthetic principles the use of technology will likely be discouraged. Because technological rationality is no longer considered to epitomize reason, persons do not have to endure criticism because they refuse to agree with technocrats. Often divergent value bases must be integrated before the adoption of a particular technology makes any sense. For example, when the beauty of the land is valued, strip mining does not have to continue because this is the least expensive method of extracting coal. As opposed to truth, persons are not obliged to accept the propositions set forth by technocrats. The responsible use of technology requires that its value base be critically examined, in order to ascertain when the adoption of technological rationality is relevant.

5. Before a free and open exchange of ideas is likely to occur, a specific approach must be taken to portraying how divergent viewpoints are reconciled. Knowledge is typically described to form a coherent system. The core of this structure, furthermore, serves as the reference point around which data are organized. Information is rejected if it contradicts the norms thought to be at the center of the system. To avoid this sort of reductionism, the organization of knowledge must assume the form of a "systase," as defined by Jean Gebser (1986, pp. 309–310). Because a systase has no center, different types of knowledge are juxtaposed to one another. Knowledge bases are thus *directly* integrated. The proliferation of value differences is encouraged without the fear of anarchy. Félix Guattari and Gilles Deleuze (1983, pp. 10–20) believe that a nonoppressive or flexible order can best be characterized as a "rhizome." Their point is that traditional Western metaphysics must be abandoned, because knowledge is believed to extend from a single, inviolable reality. This new social imagery will facilitate the "loosely coupled" organizations Karl Weick (1979) contends are imperative for controlling technology. Bureaucracies are too inflexible and insensitive, he believes, to respond to the

problems that may result from technological disasters. Institutions based on the principle of integration through diversity are able to foster innovative solutions to environmental problems.

CONCLUSION

The paramount result of establishing a reflexive environment is that reality is expanded. Realism, in a manner of speaking, is no longer synonymous with fatalism. New sources of information are available, while those who were formerly disenfranchised are able to participate in the political process. As a result, novel guidelines are revealed for solving problems, along with an entirely novel array of solutions. In order to sustain this creativity, however, reflexivity must be built into the process of institutionalizing knowledge. Knowledge bases must not be permitted to become reified, so as to guarantee that every option is given credence.

Pivotal to this essay is that increasing the use of technology will not necessarily solve the difficulties it generates. William Leiss (1979), for example, calls this approach to dealing with the consequences of technology a "technological fix." Although this pronouncement is neither new nor startling, his finding is significant in that as a result of the technological worldview this remedy often sounds feasible. Therefore, citizens should be raising the following question: Are the assumptions inherent to this worldview applicable to conceptualizing the environment? Addressing this issue reflexively, accordingly, will culminate in illustrating the limited validity of technological rationality, and facilitate the inclusion of moral, ethical, and aesthetic values in discussions pertaining to environmental planning. Hence the current propensity for using an increasing number of technological fixes can be tempered.

REFERENCES

1. Berger, Peter L. and Berger, Brigitte, and Kellner, Hansfried: *The Homeless Mind.* New York, Random House, 1973.
2. Berger, Peter L. and Luckmann, Thomas: *The Social Construction of Reality.* New York, Doubleday, 1966.
3. Boden, Margaret: *Artificial Intelligence and Natural Man,* New York, Basic Books, 1977.
4. Bolter, J. David: *Turing's Man: Western Culture in the Computer Age.* Chapel Hill, University of North Carolina Press, 1984.

5. Deleuze, Gilles and Guattari, Félix: *On the Line.* New York, Semiotext(e), 1983.
6. Dreyfus, Hubert L.: *What Computers Can't Do.* New York, Harper and Row, 1979.
7. Dreyfus, Hubert L. and Dreyfus, Stuart E.: *Mind over Machine.* New York, Free Press, 1986.
8. Ellul, Jacques: *The Technological Society.* New York, Random House, 1964.
9. Gebser, Jean: *The Ever-Present Origin.* Athens, OH, Ohio University Press, 1985.
10. Gorz, André: *Ecology as Politics.* Boston, South End Press, 1980.
11. Habermas, Jürgen: Toward a Theory of Communicative Competence. In Drietzel, Hans Peter (Ed.): *Recent Sociology,* No. 2. New York, Macmillan, pp. 115–148.
12. Habermas, Jürgen: *Theory and Practice.* Boston, Beacon Press, 1973.
13. Heidegger, Martin: *The Question of Technology and Other Essays.* New York, Harper and Row, 1977.
14. Hisker, William J.: The Promise of Technology: An Organizational Caveat. In Murphy, John W. and Pardeck, John T. (Eds.): *Technology and Human Productivity.* Westport, CT, Greenwood Press, 1986, pp. 179–189.
15. Husserl, Edmund: *The Crisis of European Sciences and Transcendental Phenomenology.* Evanston: Northwestern University Press, 1970.
16. Ihde, Don: The Historical-Ontological Priority of Technology over Science. Paper presented at the International Conference on Philosophy and Science in Phenomenological Perspective. Buffalo, NY, March, 1982.
17. Lacan, Jacques: *Ecrits.* New York, Norton, 1977.
18. Leiss, William: *The Domination of Nature.* Boston, Beacon Press, 1974.
19. Leiss, William: *The Limits to Satisfaction.* Toronto, University of Toronto Press, 1976.
20. Leiss, William: Political Aspects of Environmental Issues. In Leiss, William (Ed.): *Ecology versus Politics in Canada.* Toronto, University of Toronto Press, 1979, pp. 256–279.
21. Lyotard, Jean-Francois: *The Postmodern Condition: A Report on Knowledge.* Minneapolis, University of Minnesota Press, 1984.
22. Marcuse, Herbert: *One-Dimensional Man.* Boston, Beacon Press, 1964.
23. Melville, Stephen W.: *Philosophy Beside Itself.* Minneapolis, University of Minnesota Press, 1986.
24. Minsky, Marvin: A Framework for Representing Knowledge. In Haugeland, John (Ed.): *Mind Design.* Montgomery, VT, Bradford Books, 1981, pp. 563–568.
25. Minsky, Marvin and Papert, Seymour: *Artificial Intelligence Laboratory Memo.* Cambridge, MA, M.I.T., 1971.
26. Mumford, Lewis: *Technics and Civilization.* New York, Harcourt, Brace, and World, 1963.
27. Murphy, John W.: A Modern View on the Transfer of Technology. *Science and Public Policy,* 12(3): 144–148, 1985.
28. Murphy, John W., Mickunas, Algis, and Pilotta, Joseph J. (Eds.): *The Underside of High-tech.* Westport, CT, Greenwood Press, 1986.
29. Murphy, John W. and Pardeck, John T. Technologically Mediated Therapy: A Critique. *Social Casework,* 67(10): 605–612, 1986.

30. Murphy, John W. and Pilotta, Joseph J.: Community-based Evaluation for Criminal Justice Planning. *Social Service Review,* 57(3): 415–476, 1983.
31. Schuurman, Egbert: *Technology and the Future: A Philosophical Challenge.* Toronto, Wedge, 1980.
32. Weber, Max: *Economy and Society,* Vol. I. Los Angeles, University of California Press, 1979.
33. Weick, Karl: *The Social Psychology of Organizing.* Reading, MA, Addison-Wesley, 1979.
34. Weizenbaum, Joseph: *Computer Power and Human Reason.* New York, W.H. Freeman, 1976.

AUTHOR INDEX

L

M

T

V

W

Y

Z

SUBJECT INDEX

G

H

I

K

R

S

T